Environmental Humanities: Transformation, Governance, Ethics, Law

Volume 6

Sustainability, i.e. the demand for long-term and globally practicable lifestyles and economies, is increasingly being understood as the central challenge of our time. But just as today science is often simply equated with natural science, many people think only of the natural sciences when it comes to sustainability science. Undoubtedly, natural scientific and technical knowledge of problem relationships in dealing with nature, resources and climate is important. However, technical change does not happen on its own. In addition, the ecological challenges are simply too great not to aim for a behavioural change as well as technology. This is the starting point of this series of publications. Some questions are, for example, the conditions for individual and social change, the means or governance instruments and normative (ethical and legal) issues about the ultimate goals to be pursued. Transdisciplinary approaches should play a special role, i.e. approaches that do not operate from disciplinary boundaries but from questions of content without excessive subordination to established disciplinary dogmas. It is important to the editors that the present series stands for pluralism and expressly gives room to uncomfortable, unexpected and heterodox views and methods. In times in which sustainability research in particular (also) is increasingly influenced by the interests of clients, such openness seems necessary in the interest of truly finding knowledge.

Christian Bux

Strategies for a Sustainable and Environmentally Friendly Hospital Foodservice Management

Springer

Christian Bux
Department of Economics
University of Foggia
Foggia, Italy

ISSN 2524-5708 ISSN 2524-5716 (electronic)
Environmental Humanities: Transformation, Governance, Ethics, Law
ISBN 978-3-031-82049-6 ISBN 978-3-031-82050-2 (eBook)
https://doi.org/10.1007/978-3-031-82050-2

This Springer imprint is published by the registered company Springer Nature Switzerland AG
The registered company address is: Gewerbestrasse 11, 6330 Cham, Switzerland

Introduction

Sustainable catering strives to reduce the impact of foodservice on the environment by serving nutritious, safe and environmentally friendly meals. Italy has made a commitment to environmentally friendly food and minimizing food waste with the introduction of the minimum criteria. However, there can be a tension between these two measures, as environmentally friendly diets can lead to more waste. This handbook has the overall aim of improving the environmental, economic, and social sustainability of the food supply by considering the important variables of food waste measurement, energy efficiency, environmental impact, waste management, customer satisfaction, and consumer nutrition. The handbook reports on the results of various studies carried out in the field of hospital catering. The first study combines: (a) the assessment of energy consumption based on primary data collected in a cooking center with a production capacity of 1590 meals per production cycle; (b) the study of the consumption behavior of 984 patients in nine different hospital departments in southern Italy; and (c) the measurement of food waste based on questionnaire data. The second analysis reports the results of a three-year comparative assessment of carbon footprint and food waste measurement based on data collected through questionnaires from 3454 patients in nine hospitals, highlighting the link between sustainable choices and customer satisfaction. In terms of energy efficiency, it is possible to reduce diesel consumption by 42% and electricity consumption by 93% by switching from cook-chill to cook-hold catering, while food waste could be reduced by 85%. On average, food waste at lunch is between 8.6% and 13.9%, with the highest value being 24.6%, whereas food waste at dinner accounts from 7.0% to 18.4%, with the highest peak at 30.7%. The most wasted foods are eggs, dairy products and pasta. The weekly carbon footprint is 6.9–11.5 $kgCO_2eq$ at lunch and 6.4–7.0 $kgCO_2eq$ at dinner. One of the most original findings shows that as the rate of food waste increases, the rate of greenhouse gas emissions decreases, suggesting that the least preferred foods tend to have the lowest impact (e.g., vegetables, rice), while those with a higher consumption rate tend to have a higher level of emissions (e.g., beef meat). Nonetheless, there are foods that are both sustainable and popular with consumers, such as cod or plaice. Sustainable foodservices should go in this direction.

Contents

List of Figures

List of Tables

Chapter 1
Food Waste Measurement to Enhance Environmentally Friendly Food Consumption Patterns

Abstract Food waste measurement and monitoring over time represent a global challenge. Food waste is estimated to reach one-third of food produced for human consumption each year, which is equal to about 1.3 billion tons, and its minimization requires humans, single companies, local authorities, or entire countries to act in a sustainable manner. From a social perspective, food waste has a significant impact on hunger and malnutrition, as well as on the economy and the environment. According to recent statistics, food waste is responsible for around 6% of the entire greenhouse gas emissions, distinguishing between emissions associated with supply chain disruptions and consumer waste. Considering the impact of the Commission Delegated Decision (EU) 2019/1597 of May 3, 2019, the first chapter outlines the most commonly employed methodologies for assessing food waste quantities and composition. Specifically, it defines direct measurement tools, food diaries, the mass-balance approach (as well as the useful material flow analysis), questionnaires and surveys, as well as the waste composition analysis. The present chapter conducts a systematic literature review to examine diverse academic articles with the objective of influencing literature trends and generating empirical guidelines for academics and researchers in the field.

Keywords Direct measurement · Food diaries · Material flow analysis · Waste composition analysis · Waste measurement

1.1 Introduction

In recent years, the interest in food loss and food waste reduction strategies spread, and their health, environmental, economic and social consequences were widely investigated in literature. By definition, although there is no common understanding among researchers and practitioners in the agri-food chain, food loss could be generally intended as "the decrease in edible food mass at the production, post-harvest and processing stages of the food chain", whereas food waste is identified as "the discard of edible foods at the retail and consumer levels" (FAO 2011, 2014). However,

C. Bux, *Strategies for a Sustainable and Environmentally Friendly Hospital Foodservice Management*, Environmental Humanities: Transformation, Governance, Ethics, Law 6,
https://doi.org/10.1007/978-3-031-82050-2_1

academics and public authorities usually refer to same terms differently, also introducing possible alternative to food loss and waste such as "avoidable food waste", "unavoidable food waste" or "potentially avoidable food waste" (FUSIONS 2014). The use of different perimeters and definitions has a significant impact on policies and strategies toward their reduction, as well as on the methods that can be used for their measurement. Recently, authors and authorities tried to harmonize the current definition of food loss and waste at least in Europe, by commonly referring to food waste as "any food, and inedible parts of food, removed from the food supply chain to be recovered or disposed, (including composted, crops ploughed in/not harvested, anaerobic digestion, bio-energy production, co-generation, incineration, disposal to sewer, landfill or discarded to sea)" (FUSIONS 2016). Similar trends, which though split between food loss and food waste, are identified by the Food and Agriculture Organization of the United Nations (FAO), which states that: (i) food loss is "a decrease in mass (dry matter) or nutritional value (quality) of food that was originally intended for human consumption. These losses are mainly caused by inefficiencies in the food supply chains, such as poor infrastructure and logistics, lack of technology, insufficient skills, knowledge and management capacity of supply chain actors, and lack of access to markets. In addition, natural disasters play a role"; and (ii) food waste is "food appropriate for human consumption being discarded, whether or not after it is kept beyond its expiry date or left to spoil. Often this is because food has spoiled but it can be for other reasons such as oversupply due to markets, or individual consumer shopping/eating habits" (FUSIONS 2016). Both terms could be included in the so-called "food wastage".

In 2015, through the introduction of the Sustainable Development Goals (SDGs) by the United Nations, food waste increased its importance on the global scale. The concept of sustainability, although topical, is rather complex (Ekardt 2024). Several definitions were provided in literature. The simplest definition of sustainability is "the capability of being maintained at a certain rate or level" (Gruen et al. 2008), and researchers use the term "sustainment to denote the continued use of an innovation in practice" (Aarons et al. 2011). More complex definitions associate the term sustainability with that of benefits, health, and organizations, as follows: (i) "sustainability of organizational innovations can be thought of as the point at which new ways of working become the norm and the underlying systems and ways of working become transformed in support" (Greenhalgh et al. 2004); (ii) "the extent to which desired health benefits are maintained and improved upon over time after initial funding or supports have been withdrawn"; (3) "the extent, nature, and impact of modifications to the core and adaptable/peripheral elements of the program or innovation"; or (4) "the continued capacity to function at the required level to maintain the desired benefits" (Stirman et al. 2012). Sustainability is a comprehensive and holistic concept that refers to several dimensions, namely "time", "continued delivery", "behavior change", "evolution/adaptation" and "continued benefit" (Moore et al. 2017), but it cannot be dissociated from that of "growth", and for this reason it was introduced the topical concept of sustainable development, intended as "meeting the needs of the present without compromising the ability of future generations to meet their own needs" (United Nations 1987). From an ecological standpoint, sustainable development requires that harvest rates are equal to regeneration rates of natural resources, and that waste emission rates are equal to the natural assimilative capacities of the ecosystem into which the waste streams are released (Daly 1990).

Considering the SDGs, "Goal 12: Responsible Consumption and Production" aims at ensuring sustainable consumption and production patterns (United Nations 2024), and its sub-indicators 12.3.1A defined "Food Loss Index (FLI)" and 12.3.1B defined "Food Waste Index (FWI)" respond to the urgent need to "halve per capita global food waste at the retail and consumer levels" and to "reduce food losses along production and supply chains, including post-harvest losses" (One Planet Network 2024; United Nations 2024). FLI measures losses during the upstream stages of the food supply chain, namely post-harvest losses on farm up to but not including distribution. Based on latest data (year = 2020), food loss percentage is on average 13.3% worldwide, recording the highest peak in Sub-Saharan Africa (21.4%) and the lowest in Europe and Northern America (9.9%). Compared to the base year 2015 (FLI = 100), the average FLI increased to 101.2 at the global scale, with a significant increase in Europe and Northern America (FLI = 106.5). FWI quantities food waste at distribution and consumer level, either at foodservice or household, considering total food waste rather than single food commodities (One Planet Network 2024).

Latest statistics, viz. Food Waste Index Report 2024 (United Nations Environment Programme 2024), are based on a significant database composed of 194 datapoints across 93 countries and reveal that food waste amounts to about 115 kg per person per year in high-income countries, whereas to 88 and 86 kg in upper-middle income and lower-middle income countries, although data are insufficient. Compared to the previous concepts, FWI is based on an additional definition, that describes food waste as "food and the associated inedible parts removed from the human food supply chain" in retail, foodservice and households, including either edible or inedible parts of food. In Europe, it is possible to distinguishing between: (i) 36% at the pre-harvest stage; (ii) 10.91% at post-harvest stage; (iii) 12.36% at processing and industrial transformation stage; (iv) 6.91% at distribution stage; and (v) 33.82% at the consumer's level, but these amounts tend to vary according to the geographical area object of the analysis. In Sub-Saharan Africa, food waste at the consumer level is about 3.82%, whereas in Northern America and Oceania it rises to 39.07% (Gustavsson et al. 2011a, b).

In 2015, through the proposals of the "Closing the loop—An EU action plan for the Circular Economy", the European Commission (2015) introduced the monitoring framework for the circular economy, distinguishing among material consumption and waste generation in the field of production and consumption indices (Eurostat 2022a, b). Specifically, food waste is explicitly mentioned as one of the 10 circular economy indicators, in addition to the FLI and the FWI, and its amount was estimated at 131 kg per capita in 2021.

Food waste quantities and spatiotemporal characteristics are essential for designing reduction strategies, but their measurement represents a hard challenge, due to the lack of consistent datasets, divergent (and not standardized) methodologies, and conflicting definitions. In the absence of harmonized and uniform measurement techniques, which are essential for gathering trustworthy data and devising feasible strategies, it is virtually unattainable to attain healthy nutritional food systems, food safety, and food security. It is defined as a situation when "all people, at all times, have physical and economic access to sufficient, safe and nutritious food that meets their dietary needs and food preferences for an active and healthy life" (Committee on World Food Security 2021). Consequently, the European Commission introduced the so-called Commission Delegated Decision (EU) 2019/1597 of May 2019, supplementing Directive 2008/98/EC of the European Parliament and of the Council, on "a common methodology and minimum quality requirements for the uniform measurement of levels of food waste", with the scope of measuring food waste through a practically applicable methodology, improve the precision of monitoring tools and enable a clear and transparent indication of the amounts of food waste generated at each stage of the supply chain (Official Journal of the European Union 2019). Specifically, the measurement framework considers: (a) primary production; (b) processing and manufacturing; (c) retail and other distribution of food; (d) restaurants and food-service; and (e) households, and food waste flows shall be measured in metric tons of fresh mass. The Commission Delegated Decision (EU) 2019/1597 (Official Journal of the European Union 2019), in the Annex III on "methodology for the in-depth measurement of food waste", identifies several methods by distinguishing among those based on direct asses to food waste and other methods. In terms of methods based on direct access, also defined as direct measurement methods, the subsequent are identified: (i) weighing or volumetric assessment, through a measuring device that determines the mass of samples of food waste or fractions of total waste; (ii) scanning or counting, which consist in the assessment of the number of items that make up food waste to determine their mass; (iii) waste composition analysis, that is the physical separation of food waste from other fractions; and (iv) diaries, which require individuals or groups of individuals (e.g., families) to keep a record or log of food waste information on a regular basis (e.g., daily, weekly, monthly). Other measurement methods cannot rely on a direct access to food waste and are: (i) mass balance, which is based on an input–output analysis in a defined system and is mainly addressed through the material flow analysis; and (ii) coefficients or percentages from questionnaires and interviews, or from production statistics.

In light of these premises, the present chapter detects benefits and shortcomings of different food waste measurement methods according to a systematic literature review, focusing on those suggested by the Commission Delegated Decision (EU) 2019/1597 (Official Journal of the European Union 2019). Through a critical analysis of previous studies on the topic, the chapter intends to provide practical and theoretical information on different measurement methods and develop strategies for either public authorities or practitioners (e.g., food operators, managers, consumers) toward food waste minimization. Section 1.2 describes the materials and methods used in the chapter, Sect. 1.3 presents the bibliographic results (i.e., manifest data), whereas Sect. 1.4 presents a critical analysis of the measurement methods (i.e., latent data), with a focus on the material flow analysis. Last, the conclusions provide a synthesis on the topic.

1.2 Materials and Methods

The literature review is a useful method for building an original research based on existing knowledge, such as research papers, working papers, reports, and statistics. It is essential to carry out such a method accurately. However, researchers may face some challenges in conducting a literature review, due to the remarkable but still fragmented speed of business research in recent years (Snyder 2019), which makes it difficult to keep up with the state-of-the-art and collect significant studies on a topic. In general, researchers have the option to employ various strategies, standards, and guidelines to conduct a literature review, contingent upon the objective and scope of their research. According to previous studies, there are three different methods for carrying out a literature review, namely: (i) a systematic review, which is a technique that synthesizes research findings in a systematic, transparent, and reproducible manner (Davis et al. 2014); (ii) a semi-systematic review, which is also defined as a narrative review, which mainly investigates how a topic has developed in time across different research traditions, but does not evaluate all studies published on a specific topic (McColl-Kennedy et al. 2017); and (iii) an integrative review, also known as critical review, which aims to assess and synthesize previous studies on a topic in order to develop original theoretical frameworks and novel perspectives. They do not search for all studies published on a single topic (Torraco 2005).

To this extent, the chapter conducts a systematic literature review on food waste measurement methodologies as defined by the Commission Delegated Decision (EU) 2019/1597 on "a common methodology and minimum quality requirements for the uniform measurement of levels of food waste", viz. diaries, direct measurement, mass balance, questionnaire and surveys and interviews, and waste composition analysis. Based on the approach suggested by Snyder (2019), the review follows a stepwise approach, as follows: (i) design; (ii) conduct; (iii) analysis; and (iv) structuring and writing the review, to provide an explicit, transparent and reproducible analysis. Specifically, the research applies the PRISMA standard guidelines (Page et al. 2021). The first step regards the definition of the research items (i.e., definition of the aims and scope of the research), then the identification of the research strings related to food waste measurement methods; soon after these steps, the core part of the literature review regards the data collection and data synthesis, which is concluded by the data analysis according to a qualitative content method. The PRISMA guidelines consider different steps, such as the identification of previous literature on the topic, the removal of duplicates, the screening and the inclusion of selected studies (Page et al. 2021).

1.2.1 Database and Research Strings

By searching for food waste measurement and quantification in authoritative research database, such as Scopus, Web of Science (WoS) or Google Scholar, a plethora or studies outcome. Some of them refer to food waste drivers, some others aim at developing reduction strategies through prevention, reuse or valorization, by specifically referring to agricultural production, industrial transformation, distribution and final consumption, either at foodservice or households. However, these studies do not refer to measurement and quantification of food waste "in metric tons of fresh mass", as requested by the Commission Delegated Decision (EU) 2019/1597, but rather investigate consumer behavior, qualitative strategies or food waste drivers. Moreover, in the field of quantitative studies, most of them apply mixed methods and, as a critical issue, refer to opposing definitions or blurred system boundaries, making complex the assessment of food waste quantities. Nowadays, data collection is still scarce and uncertain, and primary data are usually integrated with secondary and third ones (Corrado et al. 2019).

In light of the systematic review standard guidelines, three research items were developed with regards to food waste measurement studies, as follows:

1. Publication timeline, geographical area, food supply chain stage and publishing journals
2. Consistency of the measurement methods applied by the researchers with the guidelines of the Commission Delegated Decision (EU) 2019/1597
3. Characteristics, shortcomings, opportunities and challenges of the selected studies toward a uniform and harmonized measurement of food waste

Considering the high variability of food waste concepts, in the chapter the subsequent definition of food waste was considered: "the masses of food lost or wasted in the part of food chains leading to edible products going to human consumption" (Gustavsson et al. 2011b), which means that agricultural material referred to article 2(1)(f) of Directive 2008/98/EC and animal by-products referred to article 2(2)(b) of Directive 2008/98/EC are not considered as food waste, as requested by the Commission Delegated Decision (EU) 2019/1597 [article 1(4)(a, b)]. In terms of inclusion criteria, the review considers quantitative academic peer-reviewed articles, whereas other studies whose focus is not on the measurement of food waste were excluded, such as studies on food waste valorization pathways or food waste drivers among consumers. In addition, books, book chapters, conference papers or review articles, as well as studies written in other languages than English, were not included.

By considering WoS Core Collection, which represent one of the most trusted, source-neutral abstract and citation database for scientific studies on the global scale, the review reflects the subsequent timeline: 2000–2020, but it seems interesting to note that one of the first studies emerging on food waste measurement dates back to 1974 (Hall et al. 2009), whereas one of the first definitions of food waste was released in Europe in 1975 by the Council Directive 75/442/EEC of 15 July 1975 on "waste". By definition, the directive defines waste as "any substance or object which the holder disposes of or is required to dispose of pursuant to the provision of national law in force", but it excludes, as was then taken up by current legislation, "animal carcases and the following agricultural waste" (Official Journal of the European Communities 1975). In terms of geographical boundaries, the research considers all countries without any exception, being food waste a global concern and not a local one. Requiring the PRISMA standard guidelines the identification of systematic research strings, the review is based on the subsequent combination of keywords, which were searched according to a TITLE-ABS-KEY query: “food waste” OR “food loss” AND “measure” OR “measurement” OR “quantification” OR “quantity” OR “quantifying” OR “method” OR “methodology” OR “estimation” OR “assessment” OR “diaries” OR “direct measurement” OR “interview” OR “questionnaire” OR “survey” OR “mass balance” OR “material flow analysis” OR “waste composition analysis”, for an amount of 34 different research strings. Soon after the identification of outcoming studies, the removal of duplicates and the screening of the selected studies (i.e., studies in line with the purpose of the review), the ultimate papers were catalogued in Microsoft Excel by highlighting the subsequent items: (i) authors; (ii) title; (iii) journal; (iv) publishing year; (v) food waste measurement method; (vi) geographical area. To these papers, others were added on existing knowledge, experts’ suggestions and serendipity.

1.3 Timeline, Journals and Geographical Area (Manifest Data)

Since several studies used mixed quantitative and qualitative methods, and only a few presented a practical or empirical measurement of food waste quantities, the selection of suitable papers for the systematic review was rather complex. Considering that the Commission Delegated Decision (EU) 2019/1597 suggests qualitative instruments for measuring food waste, such as questionnaires, surveys and interviews, several studies applied these methods, but their focus was far from the purpose of the current chapter (i.e., food waste drivers, consumers’ perception of food waste, managers’ awareness of food waste reduction strategies, farmers’ practices for biowaste valorization, etc.). To get an idea of the selection process, below are some numbers: (i) the first search, on the basis of the selected research strings, provided a number of 16,854 contributions in terms of “food waste” and 31,674 as regards “food loss”, with the highest peaks in combination with “method”, “assessment”, “questionnaire” and “survey” and for an entire amount of 48,528 contributions (considering a timeline from 1985 to 2019, the keywords “food waste” and “food loss” provide an amount of over 137,902 contributions in the section “topic” and 4508 in the “title”); (ii) soon after a TITLE-ABS-KEY query, and considering the specific purpose of the research as well as the inclusion and exclusion criteria of the review, a set of 315 contributions was created; (iii) after screening, only 58 papers were selected for final review, whereas 29 where not in line with the research.

In terms of publishing timeline, the research considered 2000–2020 as a suitable range, but the first empirical contribution on food waste measurement dates back to 2010. Not by chance, the first report on the assessment of food waste quantities was released in 2011 with the title “Global Food Losses and Waste. Extent, Causes,

Prevention" (Gustavsson et al. 2011b), highlighting the strong nexus between scientific research and legislative/authorities advances in the field, whereas no contributions were selected from 2001 to 2009. The highest quota of contributions were published from 2017 to 2019, almost certainly in the wake of the presentation of the SDGs by the United Nations in 2015 and soon after the introduction of the European monitoring framework of circular economy. In addition, the highest quota in the reference period was registered in 2019 (31%, 18 contributions), exactly when the Commission Delegated Decision (EU) 2019/1597 was released.

Although some interesting contributions published on international conference proceedings emerged from the query (Mandasari 2017; Ganglbauer et al. 2015; Le et al. 2015), only scientific articles published on peer-reviewed journals were selected. In concerning publishing journals, eight journals circulated more than one article on food waste measurement, namely: (i) Waste Management (Elsevier, online ISSN: 1879-2456), with 13 contributions (22%); (ii) Resources, Conservation and Recycling (Elsevier, online ISSN: 1879-0658), with 10 contributions (17%); (iii) Journal of Cleaner Production (Elsevier, online ISSN: 1879-1786), with eight articles (14%); (iv) British Food Journal (Emerald Insight, ISSN: 0007-070X), with five articles (8%); and (v) Sustainability (MDPI, ISSN:2071-1050); (vi) Waste Management & Research: The Journal for a Sustainable Circular Economy (Sage Journals, online ISSN: 1096-3669), with three contributions (5%); (vii) PLOS ONE, with two contributions (3%); and (viii) Food Policy (Elsevier, online ISSN: 1873-5657), with two contributions (3%). Other journals, with one contribution each, are: Science of the Total Environment (Elsevier, online ISSN: 1879-1026), Journal of Environmental Management (Elsevier, online ISSN: 1095-8630) or International Journal of Consumer Studies (Wiley, online ISSN: 1470-6431), among others.

The analysis of contributions from a geographical perspective is consistent with food waste quantities at the global scale: where it is greater, the greatest number of studies is conducted. Precisely, the vast majority of contributions was carried out in the United States (10 contributions, distinguishing between California, Colorado, Hawaii, Iowa, Missouri and North Carolina) and in Europe, with specific regard to Finland, Germany, Italy and Sweden, that account for six contributions per country. In Europe, based on the latest food waste statistics in 2020 (Miśniakiewicz et al. 2024; Eurostat 2022a, b), food waste amounts to about 58.4 million tons (Mt) of fresh mass, of which about 53% at households and 10% at foodservice. Germany, France, and Italy represent the third top producers of food waste in Europe, accounting for 10.9 Mt, 8.7 Mt and 8.3 Mt, respectively. Sweden and Finland, on the contrary, seem to be among the most virtuous countries in Europe, recording about 0.9 and 0.7 Mt of food waste. Impressive is the quota of food waste recorded at household and foodservice in Italy, namely 79%, that is one the highest in Europe. Interesting contributions investigate food waste quantities also in China, Pakistan, Saudi Arabia, Uganda, South Africa, Perú, Ecuador and Canada. However, not all studies measure food in a single country, but rather compare more than one region.

1.4 Food Waste Measurement Methods (Latent Data)

The current section presents the breakdown of food waste measurement methods by distinguishing, in order of appearance in the chapter, among questionnaire, surveys and interviews, direct measurement, diaries, waste analysis composition and mass balance (with a specific focus on the material flow analysis). Specifically, on the basis of the contributions selected for review and considering that some of them adopted mixed methods for measuring food waste, it results that: (i) 32.3% are based on questionnaire, surveys and interviews; (ii) 25.8% on direct measurement (i.e., weighing, counting, scanning, etc.); (iii) 12.9% on diaries; (iv) 6.5% on waste analysis composition; and (v) 3.2% on mass balance. In terms of food supply chain stage, it results that 35.8% of contributions investigate household food waste, 22.4% foodservice, 14.9% agricultural production (including harvest and post-harvest activities), 11.9% distribution and 7.5% the entire food chain, from planting, harvesting and transformation to end-users.

1.4.1 Questionnaires, Surveys, and Interviews

Questionnaires, surveys, and interviews are widely utilized by researchers to assess food waste, and numerous articles have utilized these tools for a quantitative assessment. A questionnaire is "a formal, structured way to collect quantitative or qualitative data from participants". It could be used to obtain food waste amounts, figures, and other additional information from producers, processors, retailers, and consumers (Amicarelli and Bux 2020; Møller et al. 2014). In general, questionnaire, surveys and interviews are treated as synonyms (hereafter questionnaire) and can be divided between those used to check existing data, in order to assess their reliability, and those used to estimate new data (Commission of Environmental Cooperation 2021). Considering the first typology (i.e., questionnaire used to collate existing data), this method is actually cost-effective and can standardize the information requested from participants, but it relies on third party data, which may affect its reliability and consistency, it can be complex to extract the precise information required (as to guarantee standardization among results) and it faces granularity troubles (e.g., different level of information among participants). In terms of questionnaires used to measure new data, this method benefits from several strengths, such as: (a) cost-effectiveness in terms of administration; (b) possibility of collecting in-depth data in terms of single food commodities or specific food supply chain stage; (c) chance of providing sociodemographic data of food wasters; and (d) opportunity of including also quantitative information, such as food waste drivers for defining feasible (and consistent) reduction strategies. However, questionnaires face significant limitations, such as respondents who tend to underestimate the amount of food waste, or respondents who do not even know the meaning of food waste, therefore producing unreliable and inconsistent data. Considering a stepwise approach, it is important to: (i) set hypotheses and determine the questionnaire approach; (ii) identify the methods by which the survey is to be administered, viz. by e-mail, by telephone, by electronic devices or face-to-face; (iii) identify respondents audience; (iv) prepare questions to quantify food waste on the basis of previous literature on the topic, considering a scientific committee and respecting the ethical standards of a committee; (v) test the survey and revise; (vi) administer the survey; and (vii) analyze data (Commission of Environmental Cooperation 2021; FLW Protocol 2016).

Contributions measuring food waste through the questionnaire method are presented in chronological order, starting from the least to the most recent, and are clustered according to the main sector of interest (i.e., primary production, distribution, foodservice, household consumption).

Willersin et al. (2015) investigate the potato supply chain in Switzerland by relying on survey data collected among three wholesalers of fresh potatoes, two wholesalers of processed potatoes, four industries of industrial transformation, five distributors and 704 households. Considering harvesting, post-harvesting, processing, distribution and consumption as system boundaries, the researchers adopted a standardized 8-page questionnaire via e-mail that included questions on either fresh mass data, ecological information and economic costs, which resulted in: (i) 15–24% of food waste at the agricultural stage; (ii) 12–24% at primary distribution (i.e., wholesale); (iii) 1–3% at secondary distribution (i.e., retail), and (iv) 15% at private households. Tostivint et al. (2016) measure food waste at primary production, processing and distribution in the diary supply chain in Pakistan, collecting data among 18 farmers, eight distributors and three experts in the field through face-to-face interviewers, phone calls and focus groups with consumers. Such a significant variety of respondents (e.g., farmers, commercial managers, academics or individual researchers) and methods underlines the high flexibility of the questionnaire for data collection, but also suggests a certain subjectivity and a great variability in the estimates, that are about 7000 t of milk waste (in strict sense) per year. Specifically, primary and secondary data come from multinationals operating in Pakistan, such as Nestlé. On the same trends, Hartikainen et al. (2018) quantifies food waste amounts at primary production in Nordic countries (i.e., Finland, Sweden, Norway and Denmark), disseminating 2500 final questionnaires among food producers and farmers soon after a pilot research among 21 respondents. Quantitatively, food waste in Nordic countries amounts to 0.8 Mt per year, to which additional 0.1 Mt should be added for the rearing of animals (for an amount of about 0.9 Mt). However, changing system boundaries according to a different food waste definition reduces the estimates of food waste to about 0.3 Mt (− 63%), because inedible parts of food (e.g., skins, peels, bones), as well as food addressed to animal consumption as feed, are not accounted. The research introduces a significant concept, defined as "side flow", which is divided in: (a) "food-chain-suitable-side-flow", viz. food that is still expected to be consumed by humans, but then is discarded; (b) "inedible side flow", that contains contaminated food products due to diseased or pests, for instance; (c) "valorization and conversion" flows, intending food waste used as feed or recovered into bio-bases materials; and (d) "left in field, composted, sent to waste treatment" flows. One similar research is conducted by Baker et al. (2019), whose purpose is to measure food waste in primary production in Northern and Central California, collecting data for twenty different crops (i.e., bunch spinach, celery, green leaf lettuce, iceberg lettuce, napa cabbage, romaine hearts, romaine lettuce, romaine tomatoes, round tomatoes, sweet corn, etc.) in 123 fields. Specifically, the research tries to overcome problems associated with uncertainties and subjectivity of field managers (defined as "the grower estimate of on-farm food loss"), since if a grower was unsure about food waste quantities, no response was recorded. Moreover, questionnaire data were integrated with quantitative methods, since the authors "estimated the loss in kilograms/hectare by multiplying the total amount of unharvested produce in the plot by the inverse of the plot's proportion of a hectare". On average, edible food products left in field after harvest were about 11 t per ha. Last, similarly to Tostivint et al. (2016), also Wesana et al. (2019) investigate one vertically integrated dairy chain in Western Uganda, from farm production to industrial transformation and distribution. Through personal observations and interviews with key managers operating in the organization, the research estimated food waste at about 1.5 t of milk, 0.3 t of plain yogurt and 0.1 t of mango yogurt. This research introduces a concept that represents the basis of the current book, namely the nutritional value embedded in food waste. By definition, food waste "translates to a substantial nutritional value disappearing from the food system, which compromises the ability of people to meet their nutrient recommendations" affecting food security (Wesana et al. 2019).

The hospitality industry is a highly intricate sector of the economy, comprising diverse units such as canteens, hospitals, hotels, schools, restaurants, and universities. However, a limited number of studies have been conducted in the industry to distinguish between individual establishments (Malefors et al. 2019). The present book, to fill in this gap, reports the findings of different studies conducted in the hospital foodservice in terms of food waste measurement and its consequences, and presents the results from original researches conducted in the field. Pirani and Arafat (2016) estimate food waste quantities in the hospitality sector in the United Arab Emirates and Abu Dhabi, through face-to-face interviews with 45 managers of either hotels or restaurants. Specifically, the research interviews engineers, environmental managers, as well as quality, food safety and hygiene managers, but any quantitative results were collected through questionnaire, since: (i) one first measurement was obtained from the kitchen staff of a catering company; and (ii) one second measurement was based on the material flow analysis, data being based on the experience of chefs and their staff. Based on these data, the research estimates also the carbon and water footprint associated with food waste, a specific topic which will be developed in the next chapters of the current book. Together with carbon and water footprint, as well as the assessment of nutritional losses, other studies can be conducted on the basis of food waste quantities, such as those related to the energetic and economic "laten (or hidden) flows". In this light, Amato and Musella (2017) measure food waste in bars and restaurants among 89 participants in the historic center of Naples (Italy), and questionnaires were anticipated by telephone calls and letters describing the research content, so as to increase reliability and completeness of results. Per year, it was estimated an amount of 744 t of waste, distinguishing between 323 t of bar waste and 421 t waste. The research provides an in-depth assessment of the different food commodities, which are wasted the most, distinguishing between ready-to-eat products (i.e., pastry products, rotisserie products, restaurant products) and raw or semi-finished products (i.e., bread, fruit and vegetables, meat products, cheese products). In China, with regards to four cities (Beijing, Shanghai,Chengdu, and Lhasa), Wang et al. (2017) measure per capita food waste according to a questionnaire method in 195 restaurants, and for each restaurant 20 tables of consumers were randomly selected (about 3900 tables were interviewed). From the methodological perspective, the authors adopted a mixed method that included either interviews with chefs (and separation of leftovers after consumption), as well as interviews with consumers (and on-site weighing) for a double-check. On average, it was estimated an amount of 93 g of food waste per capita per meal. Similarly for the previous studies, also Pflugh Prescott et al. (2019) adopted a mixed method to measure food waste in school kitchens, viz. interviews with key respondents and 80 h of structured kitchen observations. In particular, the research estimates food waste among 14 schools and district kitchens in Colorado using a stratified random sampling, with an assessment of 43.8 g per student, of which 32.5 g (i.e., three-quarters) of edible food. Last, Eriksson et al. (2020) quantify food waste in 20 hospitals located in three different regions in Sweden trough an online questionnaire. However, although this research may represent one of the few studies in hospitals in Europe, the questionnaire was sent to one contact person, and a general underestimation was perceived across the entire study. The average food waste was about 111 g per hospital patient per meal, ranging from 27 and 184 g. Daily food waste per hospital was estimated up to 239 kg.

Food waste at households represents more than half of the entire food waste generated along the food supply chain, but its estimation is rather complex due to bias associated with respondents (e.g., willingness to demonstrate "more socially acceptable" behaviors, absence of coaching, lack of awareness of the concept of food waste). Several studies carried out questionnaire to measure food waste at households, as follows: (i) Chakona and Shackleton (2017) estimated an amount of 121 g of prepared food, 268 g of unprepared food and 77 mL of drinks per households, as an average of 554 households in Richards Bay, Dundee and Karrismith (South Africa); (ii) Szabò-Bòdi et al. (2018) recorded about 68 kg of food waste per person per year (about 180 g per person per day), of which 48% still available, among 295 households in Hungary; (iii) Djekic et al. (2019) estimated an amount of 0.95 kg of food waste per person per week (about 135 g per person per day) among 801 households in Sarajevo and Banja Luka, Bosnia and Herzegovina; (iv) Elimelech et al. (2019) measured an amount of 0.92 kg per capita per week (132 g per person per day) among 187 households in Israel; (v) Giordano et al. (2018) estimated an amount of 530 g of food waste per person per day among 388 families in Italy; (vi) van Herpen et al. (2019) quantified an amount of 0.61–1.2 g of food waste per household per week (about 8–17 g per person per day) among 143 households in the Netherlands; (vii) Kasza et al. (2020) measured an amount of 65.49 kg of food waste per person per year (about 179 kg per person per day) among 165 households in Hungary.

1.4.2 Direct Measurement

By definition, direct measurement (i.e., weighing or volumetric assessment) regards the "use of measuring device to determine the mass of samples of food waste or fractions of total waste, directly or determined on the basis of volume" and "it includes measurement of separately collected waste" (Official Journal of the European Union 2019). Direct measurement could be conducted through a large variety of methods such as a direct count or a process of weighing, which return the most accurate food waste quantities but require, at the same time, expertise, time and costs. In general, among others, the main strengths of direct measurement regard accurate data, monitoring of progress over time, as well as tracking of food waste drivers, but it is time-intensive, expensive, requires direct access to waste streams and the feasible methods vary from sector to sector and according to different supply chain stages (Commission of Environmental Cooperation 2021).

In primary production, common methods for direct measurement could rely on random samples from the crop or product under-research. Hartikainen et al. (2018) measure food waste in primary production, either for agriculture, aquaculture or fisheries, by developing a mixed method composed of questionnaires, direct in-field measurement, interviews and literature data to fill in data gaps. Specifically, the research considered 42 subcategories of food in Finland, Sweden, Norway and Denmark, and carried out different case studies on six products, namely carrots, onions, cereals (i.e., wheat or rye), green peas, field peas and farmed fish (i.e., rainbow trout or char). One interesting toolkit was developed by Johnson et al. (2018), which measure food waste in primary production across 13 field of 121 ha in North Carolina (United States). This toolkit helps farmers assess the amount of marketable produce remaining in field and is based on a stepwise approach, as follows: (i) sampling of the crop area; (ii) selection and marking of rows randomly; (iii) harvesting the rows; (iv) sorting of samples into categories; (v) weighing and recording samples in each category; (vi) analysis of data and enlargement of the results to the entire field to calculate a potential estimate of the field (Commission of Environmental Cooperation 2021). Schneider et al. (2019) quantifies food waste in primary production, with regards to the potato production, through six case studies in Austria and Germany. The research is based on an on-site measurement of pre-harvest losses for potatoes by comparing two years, with an average result of 0.7 t/ha of non-harvested potatoes and about 0.9 t/ha of over-sized, damaged, and sorted out potatoes.

Direct measurement for food waste distribution at retail could refer to the electronic scanning, which are already implemented by retailers for inventory and sales. However, this is the case of packaged food with barcodes, whereas fresh food, such as fruits and vegetables, bakery products, etc. make such an assessment more complex and the development of a direct measurement for these commodities can be rather expansive and discourages organizations. Eriksson et al. (2012) measure food waste in six stores located in Sweden (i.e., Willys chain) by scanning products with barcode, whereas products without barcode were weighted and manually entered into the research database. Results highlighted an amount of food waste per year distinguishing by category, as follows: (i) 10 t of potatoes; (ii) 7.31 t of lettuce; (iii) 6.75 t of tomatoes; (iv) 5.37 t of sweet pepper; (v) 4.50 t of carrots; (vi) 4.37 t of bananas; (vii) 3.78 t of apples; (viii) 3.74 t of oranges; (ix) 3.66 t of grapes; and (x) 3.06 t of pears. Same methods and similar results were carried out by Brancoli et al. (2017) for the assessment of food waste in a mid-size urban supermarket (i.e., 410 m^2 of sale area) in Sweden, resulting in an amount of food waste of 22.5 t per year.

Seven studies were conducted on food waste through direct measurement tools in the hospitality industry, as follows: (i) Kallbekken and Saelen (2013) measure food waste in 53 hotels from Nordic Choice Hotels, recording an average amount of pre-treatment food waste of 30.90 kg per hotel and of post-treatment food waste of 31.02 kg; (ii) Costello et al. (2015) quantify food waste in four different campus dining facilities in Columbia (United States) through a direct weighing, with a result of pre-consumer food waste of 24.6 g (6.7 g inedible and 17.9 g still edible) per customer and 47.1 g (9.4 g inedible and 37.7 g still edible) of post-consumer food waste; (iii) Strotmann et al. (2017) adopt a threefold method (i.e., on-site visit, on-site measurement, off-site analysis) through electronic scales for measuring food waste in one hospital, one cafeteria and one retirement home in Germany, resulting in 25.6 ± 4.6% of food waste for hospital patients, 19.8 ± 8.3% of food waste for customers and 21.4 ± 6.7% of food waste for retired residents; (iv) Boschini et al. (2020) estimate food waste in five primary school canteens in Italy by weighing plate waste with electronic scales, with a result of 114.4 g of food waste per person, of which 107 g of still edible food and 7.4 g of inedible fraction; (v) Beretta and Hellweg (2019) adopt a physical assessment combined with staff member interviews and literature data for measuring food waste in the foodservice through 13 case studies in Switzerland, Germany, Austria, Finland and the United Kingdom. Considering that the research varies from hospitals to restaurants and school canteens, the highest food waste amount was recorded among hospital patients (302 g per meal), whereas the lowest in hotels (64 g per meal), with average values at primary schools (147 g per meal) and business canteens (189 g per meal); (vi) last, Giboreau et al. (2019) adopt a peculiar measurement method, namely a photo-assessment by a camera on a tripod from 1 m of distance and a 45° angle, in one school cafeteria with 215 children in France. Although the measurement appears novel, it allows for identifying percentages of food waste rather than a precise fresh mass amount, and it results in 66% of food waste for warm vegetables in side-dishes and 42% of food waste for cold vegetables as starters.

In the field of direct measurement methods at households, several tools can be adopted, and an effective one was introduced by Khalid et al. (2019). Specifically, researchers provided households with plastic bags for the assessment of their one-day food waste, and each food waste flow was separated from the others in other to obtain a precise amount for each food commodity, as follows: (i) 35.01 g per household per day of cooked food waste; (ii) 11.62 g of vegetables; (iii) 9.78 g of fruits; (iv) 2.45 g of cereals; (v) 1.98 g of dairy products.

1.4.3 Diaries

The Commission Delegated Decision (EU) 2019/1597 defined food diaries (also known as kitchen diaries) as follows: "an individual or group of individuals keeps a record or log of food waste information on a regular basis" (Official Journal of the European Union 2019), highlighting the importance of defining spatiotemporal characteristics of the analysis for a correct food waste assessment. Several benefits are associated with diaries, such as the opportunity to collect qualitative information like food waste drivers, but also information related to the different pathway where food waste is addressed (e.g., composting, recycling, landfilling), otherwise difficult to measure. Moreover, diaries can assume diverse forms and could be paper-based, electronic or even photographic, but among its limitations it results that diaries can be expensive (i.e., generally, participants to experimentations receive financial contributions by researchers), that participants tend to underestimate their waste flow and that the responsible for food waste diaries (e.g., head of the household) cannot perfectly know the guidelines or the rule for estimating food waste, as long as its definition. On average, some steps should be followed for conducting a food waste assessment through diaries, as follows: (i) selection of the quantification method of food waste (i.e., weighing, volumetric assessment, visual approximation); (ii) identification on how diaries should be administered to participants (e.g., printed version, electronic version, smartphone), according to the financial opportunities of the researchers, as well as to the audience of the diary (e.g., elderly people are more likely to prefer paper-based diaries instead of electronic ones, whereas young people prefer smart-app); (iii) selection of the audience, according to a random sampling strategy so as to avoid biases; (iv) recruitment of participants, which should expect a financial reward for respondents considering that participating in a diary is rather time-consuming and people need to be incentivized in order to avoid drop-out; (vi) development of the questions to be included in the diary, such as quantitative (e.g., how much food waste do you generate at lunch?) or qualitative ones (e.g., which food do you waste? how was it purchased? how was it disposed? why was it thrown away?); (vii) conduction of a pilot experiment to test the diary; (viii) administration of the diary, which represent the core step of the assessment; (ix) data analysis (FLW Protocol 2016).

Based on previous studies, participants to diaries appear rather enthusiast to participate (Langeley et al. 2010), but soon after few days some difficulties are likely to appear. First, the recruitment of participants seem hard due to high drop-out rates or quite-quitting phenomena. Secondly, there is a high risk of self-selection, since only people interested in the experiment are likely to participate. Third, although respondents are asked to personally weight food waste, there is a strong risk of approximation due to undervaluation of food waste flows. Last, so as to demonstrate a "socially acceptable" behavior, some participants tend to eat differently during the experiment period or are likely to dispose in a more sustainable manner, compromising the reliability of the results (Quested et al. 2020). Table 1.1 presents the characteristics of food diaries selected for review.

Table 1.1 Characteristics of food diaries

Reference	Period	Sample	Results
Langeley et al. (2010)	1 week	13 households in the UK	0.20 kg per person per day
Katajajuuri et al. (2014)	2 weeks	380 households in Finland	0.23 kg per person per day
Silvennoinen et al. (2014)	2 weeks	380 households in Finland	0.59 kg person per day
Ritcher and Bokelmann (2017)	1 week	25 households in Germany	N/A
Delley and Brunner (2018)	2 months	506 households in Switzerland	0.02 kg per person per day
Giordano et al. (2018)	1 week	20 households in Italy	0.07 kg per person per day
Giordano et al. (2019a)	1 month	385 households in Italy	0.07 kg per person per day
Giordano et al. (2019b)	2 months	385 households in Italy	0.07 kg per person per day
Leverenz et al. (2019)	2 months	16–37 households in Germany	0.01–0.04 kg per person per day
van Herpen et al. (2019)	1 week	143 individuals in Netherlands	0.15 kg per person per day
Ilakovac et al. (2020)	1 week	118 households in Croatia	0.21 ± 0.15 kg per person per day
Moreno et al. (2020)	1 week	489 households in the USA	0.39 kg per person per day

1.4.4 Waste Composition Analysis

By definition, the waste composition analysis is the "physical separation of food waste from other fractions in order to determine the mass of the fractions sorted out" (Official Journal of the European Union 2019) and is used by organizations (or individuals) to separate food waste streams from those that do not contain any food, for instance packaging or solid waste (FLW Protocol 2016). Considering that food waste streams tend to degrade easily, especially according to different climate areas, researchers should carry out waste composition analyses quickly. Moreover, in light of the high costs for implementing such an analysis, in generally researchers rely on a small sample size of waste flows, which makes results uncertain and therefore requires a high level of expertise in terms of sampling theory and statistics for data analysis.

Some studies were depicted from the literature search. In the field of foodservice, Rajan et al. (2017) measure food waste generated by two primary foodservice providers in the United States through a simple sorting of food waste streams from composting green bins located in the university campus. It resulted that food waste was about 1.1 t of food waste per week.

On the side of households, Hanssen et al. (2016) measure food waste streams among 12 households in Norway through an anonymous waste sampling. Specifically, waste was collected by municipalities on ordinary collection days over a seven days period by 160 L waste bags, with the results of 0.93 kg of total waste per household per day, distinguishing between: (i) 0.53 kg of food waste; (ii) 0.31 kg of still edible food waste; and (iii) 0.08 kg of packaging materials (included in the sorting). Similarly, Elimelech et al. (2018) quantify food waste among 192 households (for an amount of 634 individuals) in Israel, by sampling 1257 waste bags. From the methodological perspective, the researchers measured waste of individual households on a daily basis, and results seem in line with previous studies (0.57 kg per capita per day). van der Werf et al. (2018) measure food waste streams among 28 households in Canada, recording an amount of 0.34 kg per capita per day, which is rather lower compared to Europe and Israel, whereas van Dooren et al. (2020) quantify food waste in the Netherlands in a range from 0.08 to 0.10 kg per person per day (considering fruit, vegetable and garden waste).

1.4.5 Mass Balance Approach

In light of the Commission Delegated Decision (EU) 2019/1597, the mass balance approach represents a method for food waste measurement when researchers do not have any direct (physical) access to food waste streams and is defined as the "calculation of the amount of food waste on the basis of the mass of inputs and outputs of food into and out of the measured system, and processing and consumption of food within the system". On the basis of previous studies on the topic, such as Beretta et al. (2013), Pirani and Arafat (2016) and Caldeira et al. (2019), one of the most common methods for an input–output assessment if the material flow analysis, based on the mass-balance principle and defined as a "systematic assessment of the state and change of material flow and stock in space and time" (Brunner and Rechberger 2017) (see paragraph 1.4.5.1). Considering previous studies on the topic, Beretta et al. (2013) measure food waste streams in Switzerland, with regards to 22 different food commodities, namely: (i) fruits (i.e., apple, fresh fruit, berries, canned fruits); (ii) vegetables (i.e., potatoes, fresh, storable and processed vegetables); (iii) cereals (i.e., bread and durum wheat, rice, maize); (iv) sugar, oil and fats; (v) dairy products (i.e., milk, cheese, butter); (vi) eggs; (vii) meat (i.e., pork, poultry beef); and (viii), fish. Compared to previous studies, which focus on single or more than one supply chains stage, the current research through the material flow analysis considers the entire food value chain, distinguishing between harvest losses, conversion losses from feed to animal products, dairy/egg/meat production losses, postharvest losses, processing losses, distribution waste, foodservice waste and household waste. In terms of downstream stages, the highest food waste was recorded at households (6570 TJ), followed by foodservice (925 TJ). On the same wave, also Caldeira et al. (2019) estimate food waste in the European Union with regards to meat, fish, dairy products, eggs, cereals, fruits, vegetables, potatoes, sugar and oil crops, from primary production to household consumption, including processing, distribution and foodservice. Considering about 129.2 Mt of food waste in Europe, it is possible to distinguish between: (i) 32.3 Mt of food waste at primary production; (ii) 30.6 Mt at industrial transformation; (iii) 6.7 at distribution; (iv) 10.3 Mt at foodservice; and (v) 49.6 Mt at households. Last, although with regards to a single industry (i.e., the potato industry), Amicarelli et al. (2020) quantify food waste distinguishing between harvest, post-harvest and industrial processing waste, according to a cradle-to-gate approach.

1.4.5.1 Material Flow Analysis

Based on the definition provided in the Handbook of Material Flow Analysis (Brunner and Rechberger 2017), the material flows analysis is a "systematic assessment of the state and change of flows and stocks of materials within a system defined in space and time". It allows for the connection between sources, pathways, and the intermediate and final sinks of materials, and it is mainly based on the law of conservation of matter (i.e., principle of mass conservation), which states that matter cannot be created or destroyed in an isolated system (Sterner et al. 2011). Since during any physical of chemical change the total mass of the products remains equal to the total mass of the reactants (Ball et al. 2011), the results of the material flow analysis can be controlled comparing all inputs, stocks, and outputs of a defined process. Therefore, through a simplified mass balance, this tool represents an intuitive and appealing decision-support method in environmental, resource and waste management. Practically, the material flow analysis deals with a consistent set of information about flows and stocks over time in a defined space, either managing energy, natural resources, emissions or socioeconomic issues. It connects the anthroposphere (driven by men) with the environment (ruled by nature).

By definition, the word "material" defines either substances or goods. According to Atkins and Beran (1992), substances are "a single type of matter consisting of uniform units", while goods are "substances or mixtures of substances that have economic value assigned by markets". Goods can have either a positive value (e.g., fuel, biomass) or a negative one (e.g., solid waste), and often include also immaterial goods such as electricity. The term "process" includes agricultural operations, transport, manufacturing and storage of materials. In terms of "stock", it embodies material reservoirs generally expressed in terms of weight. Furthermore, stocks are fundamental variables of a defined system's metabolism, in order to evaluate steady-state conditions or dynamism. Subsequently, processes are linked by "flows" (i.e., mass per time) or "fluxes" (i.e., mass per time and cross section), so-called imports or exports or, better, inputs (when entering the process) or outputs (when exiting the process). The "system" includes material flows, stock and processes within defined system boundaries, and can be easily defined, for instance, as countries, regions, households, industrial plants, etc. Further definitions regard: (a) scale; (b) single processes; (c) small and complex material systems; (d) material flow account and related material flow analysis-derived indicators. In terms of "scale", it is suitable to analyze systems on all scales, from single processes (i.e., micro-level) to complex ones, even including several subsystems (i.e., macro-level). The investigation among "single processes" means beginning from the basis of the industrial process in a well-defined and transparent way, even considering small subsystems, single goods or single substances. However, it is possible to apply the material flow analysis to "complex systems", enlarging the analysis to national economies, global materials utilization or national waste management systems (Allesch and Brunner 2015). Last, Brunner and Rechberger (2017) refer to "material flow analysis-derived indicators", intending those as essential instruments to develop and implement national policies.

To sum up, the material flow analysis connects and quantifies sources, pathways and sinks of materials in physical units and the calculations are based on the law of conversation of mass (Baccini and Brunner 2012), as follows (Eq. 1.1):

$$\sum_{i=1}^{i=k} \dot{m}_{\text{input},i} = \sum_{m=1}^{m=n} \dot{m}_{\text{output},m} + \dot{m}_{\text{stock}} \quad (1.1)$$

where $\dot{m}_{\text{input},i}$ represents the input flows entering a process, whereas $\dot{m}_{\text{output},m}$ the amount of waste.

1.5 Challenges in Measuring Food Waste

The complexity of food waste, from either the conceptual or the measurement standpoint, impedes a clear and comprehensive understand of the underlying problem, and reduces the likelihood to minimize its amount at the local and the global scale (Garske et al. 2020).

One of the main challenges to adopt suitable strategies to measure and reduce food waste regards the uncertainty caused by the plethora of definitions provided in literature and the lack of existing data (Amicarelli and Bux 2020). In terms of definitions, Directive 2008/98/EC of the European Parliament and of the Council of 19 November 2008 on waste defined food waste as "all food [defined in article 2 of Regulation (EU) 178/2002] that has become waste" (Official Journal of the European Union). However, despite attempts to make the definition of food waste homogeneous also through European and international projects (e.g., FUSIONS), the definitional problem at a legislative level remains. First, a clear nexus between food waste and the waste hierarchy for food products is missing. Secondly, the absence of a quantified, measurable and sanctionable minimization target for food waste in the current version of the Waste Framework Directive (Directive 2008/98/EC) limits the possible opportunities for its reduction. Moreover, as discussed in the previous paragraphs, the choice of the measurement methods for measuring food waste among Member States remains open, making the replicability and comparability of the results across countries still complex (Garske et al. 2020). From a policy perspective,

FLI and FWI, as well as the more general reduction targets presented by the SDGs (i.e., to "halve per capita global food waste at the retail and consumer levels" and to "reduce food losses along production and supply chains, including post-harvest losses") should be supplemented by clear, ambitious, measurable and sanctionable reduction target for food waste in all Member States. Furthermore, a legal anchoring of food waste to the waste hierarchy of food products could be implemented, with specific regard to food prevention and food donation to charities (Garske et al. 2020).

In terms of food waste quantities, common data on food waste estimate its amount from 280 to 300 kg per capita per year in Europe (Food and Agriculture Organization 2011), but a study commissioned by the European Union in 2010 quantified it at about 179 kg per capita per year (Monier et al. 2010). These different results are due to different databases, different baseline years or different methods applied (i.e., primary data from diaries, direct measurement, questionnaires, etc.). In addition, Eurostat data (i.e., secondary data) are subject to large uncertainties due to different calculation methods in each Member State (Bräutigam et al. 2014). Differences in results are due also to outdated and concentrated data on specific regions or steps of the supply chain (mostly at the household level) (Bräutigam et al. 2014). Researchers face several challenges in measuring food waste along the supply chain, which encompasses numerous actors and processes, and makes the measurement activity time and economic costly (Food and Agriculture Organization 2014). However, to resolve this problem, researchers could define specific boundaries to their research and switch from aggregate to disaggregate data, with a clearer degree of detail.

1.6 Policy Instruments to Tackle Food Waste

Although food waste represents an environmental (i.e., in terms of negative impact and consumption of natural resources), economic (i.e., in terms of economic losses and disposal costs for organizations) and social (i.e., with reference to malnutrition and to malnutrition) problem, there are several shortcomings from a policy standpoint. Over time, various political statements and initiatives have been introduced, such as the circular economy and waste law, the Common Agricultural Policy (CAP) and the Common Fisheries Policy (CFP), the law of food safety and packaging, norms and date labeling as well as the relevant law for food disposal, foodsharing, and food donations. Nonetheless, there is room for improvement.

In Sweden, through the Swedish Environmental Code, it is stated that all actors must consider every possibility to reduce the amount of (food) waste generated in any facility, unless this is unreasonably expensive. In a case-study in the catering sector, Eriksson et al. (2023) explored the potential benefits of applying potential benchmarking thresholds for illegal levels of food waste and best voluntary practices that can achieve low levels of food waste. Public authorities, by forcing catering units to use best possible technologies in a broad perspective (i.e., intending best possible technologies in use at full scale, in the same or similar sector at any site on the planet), should switch from voluntary to mandatory practices. Through the adoption of sustainable planning, cooking, serving, communication, and organization activities, catering could reduce food waste from 13 to 76%.

Despite food waste reduction objectives set by the SDGs (i.e., to "halve per capita global food waste at the retail and consumer levels" and to "reduce food losses along production and supply chains, including post-harvest losses"), Garske et al. (2020) discussed the opportunity to amend the Waste Framework Directive (Directive 2008/98/EC) by taking into account particularities of specific food production, since some generate unavoidable waste while others can be easily recovered. This is the case of hospital food waste, which cannot be donated for reasons of contamination and food safety, and retail food waste, which can instead be donated. Moreover, it would be advisable to introduce a quantified, measurable, and sanctionable food waste reduction target. This represents an ambitious target toward food waste reduction and requires a harmonized revision of both the circular economy legislative framework and the specific measures introduced by the CAP regulations (Garske et al. 2020). Furthermore, public authorities should remove legal and private barriers to food waste donations and should harmonize local regulations on labeling and fiscal incentives to promote donations. For example, public authorities could prohibit the wastage of still edible food at the retail level by implementing a food donation obligation, or support foodsharing initiatives from civil society and organizations/industries (Garske et al. 2020). This is in view of a more stringent review of food waste policies.

1.7 Conclusions

Food waste measurement (and monitoring over time) represents a hard challenge, due to the lack of consistent datasets, divergent (and not standardized) methodologies and conflicting definitions. One of the overarching steps in measuring food waste represents the selection of a suitable measurement method, as outlined by the Commission Delegated Decision (EU) 2019/1597, according to the system boundaries (i.e., micro, meso or macro-level), the stage of the supply chain and the study area. In the light of the current chapter, some take-home messages can be defined, either for practitioners willing to measure food waste within their organization or for public authorities in defining guidelines for harmonized and uniform measurement methods (or indicators), as follows: (i) data availability represents a crucial point to achieve reliable and consistent data on food waste, and in its absence no clear and transparent strategies can be defined for reducing food waste amounts at primary production, industrial processing, distribution, foodservice and households; (ii) a data collection plan represents an essential step for organizations (or individuals) toward the development of sustainable and environmentally friendly production (and consumption) patterns, but specific skills among practitioners and researchers must be achieved and the role of education is essential for achieving this objective; and (iii) data must be replicable and comparable, and for this reason it is important to rely on standardized and harmonized methods, such as those suggested by the

Commission Delegated Decision (EU) 2019/1597. However, likewise the case of the material flow analysis, many technical aspects cannot be forgotten, such as the concept of food waste streams, that of functional unit or system boundaries, and the definition of the spatiotemporal characteristics of the system under-research, which represent some grounding points for a clear food waste measurement.

Studies on food waste measurement are comprehensive, and the vast majority of them focus on the downstream stages of the food supply chain, viz. foodservice and household consumption, which contribute to more than half of the global food waste. At households, food waste amounts vary from 0.01 kg per capita per day in Germany (Leverenz et al. 2019), estimated through food waste diaries, to about 0.57 kg in Israel (Elimelech et al. 2018), estimated through a waste composition analysis. On average, food waste at households is 131 kg per capita per day, and similar results were quantified by van Herpen et al. (2019) (0.131 kg), Elimelech et al. (2019) (0.132 kg) and Djekic et al. (2019) (0.136 kg) through the questionnaire method, whereas the lowest results were obtained by food diaries (from 0.01 to 0.07 kg). In the hospitality sector, with a specific focus on foodservice (i.e., canteens and restaurants), food waste quantities ranged from 0.04 kg per meal in the United States (2019) to 0.15 kg per meal in Switzerland, Germany, Austria, Finland and the United Kingdom (2019) and were estimated through the direct measurement. However, average data (0.08 kg per meal) were estimated by questionnaires, as in the case of food waste measurement at households.

More practically, researchers and practitioners involved in food waste measurement should answer the subsequent questions: "how much", "what" and "where" food waste occurs, so as to comprehend its specific quantity, its qualitative composition and the moment (i.e., the supply chain stage) of generation and to more easily define suitable reduction strategies. The subsequent chapters, on the basis of the notion of measurement, intend to define the concept of environmental impacts associated with food waste by presenting some case studies conducted in the hospital foodservice in Italy. Chapter 2 introduces the environmental consequences of food waste on the global scale and the minimum environmental criteria in Italy, whose aim is to reduce the negative consequences of food production, consumption and waste.

References

Aarons G, Hurlburt M, Horwitz S (2011) Advancing a conceptual model of evidence-based practice implementation in public service sectors. Adm Policy Ment Health Ment Health Serv Res 38:4–23. https://doi.org/10.1007/s10488-010-0327-7

Allesch A, Brunner PH (2015) Material flow analysis as a decision support tool for waste management: a literature review. J Ind Ecol 19(5):753, 764

Amato M, Musella M (2017) Quantification of food waste within food service in the historic center of Naples: a case study. Qual Access Success 18(2):22–28

Amicarelli V, Bux C (2020) Food waste measurement healthy and environmental-friendly food system: a critical review. Br Food J 123(8):2907–2935

Amicarelli V, Bux C, Lagioia G (2020) How to measure food loss and waste? A material flow analysis application. Br Food J 123(1):67–85

Atkins PW, Beran JA (1992) General chemistry. Scientific American Inc., New York

Baccini P, Brunner PH (2012) Metabolism of the anthroposphere: analysis, evaluation, design. The MIT Press, Massachusetts, United States

Baker GA, Gray LC, Harwood MJ, Osland TJ, Tooley JBC (2019) On-farm food loss in northern and central California: results of field survey measurements. Resour Conserv Recycl 149:541–549

Ball DW, Hill JW, Scott RJ (2011) The basics of general, organic, and biological chemistry. Flatworld, Boston, United States

Beretta C, Hellweg S (2019) Potential environmental benefits from food waste prevention in the food service sector. Resour Conserv Recycl 147:169–178

Beretta C, Stoessel F, Baier U, Hellweg S (2013) Quantifying food losses and the potential for reduction in Switzerland. Waste Manag 33(3):764–773

Boschini M, Falasconi L, Cicatiello C, Franco S (2020) Why the waste? A large-scale study on the causes of food waste at school canteens. J Clean Prod 246:118994

Brancoli P, Rousta K, Bolton K (2017) Life cycle assessment of supermarket food waste. Resour Conserv Recycl 118:39–46

Bräutigam KR, Jörissen J, Priefer C (2014) The extend of food waste generation across EU-27: different calculation methods and the reliability of their results. Waste Manag Res 32:683–694

Brunner PH, Rechberger H (2017) Handbook of material flow analysis: for environmental, resource and waste engineers, 2nd edn. CRC Press, Taylor & Francis Group, LLC, Boca Raton, London and New York

Caldeira C, De Laurentiis V, Corrado S, van Holsteijn F, Sala S (2019) Quantification of food waste per product group along the food supply chain in the European Union: a mass flow analysis. Resour Conserv Recycl 149:479–488

Chakona G, Shackleton CM (2017) Local setting influences the quantity of household food waste in mid-sized South African towns. PLoS ONE 12(12):0189407

Commission of Environmental Cooperation (2021) Why and how to measure food loss and waste: a practical guide—version 2.0. Commission for Environmental Cooperation, Montreal, Canada

Committee on World Food Security (2021) Global strategic framework for food security & nutrition (GSF). Access date 15 June 2024. Available from: https://www.fao.org/cfs/policy-products/onlinegsf/en/#:~:text=Food%20security%20exists%20when%20all,%2C%20access%2C%20utilization%20and%20stability

Corrado S, Caldeira C, Eriksson M, Hanssen OJ, Hauser HE, van Holsteijn F, Liu G, Oestergren K, Parry A, Secondi L, Stenmarck A, Sala S (2019) Food waste accounting methodologies: challenges, opportunities, and further advancements. Glob Food Sec 20:93–100

Costello C, Birischi E, McGarvey RG (2015) Food waste in campus dining operations: inventory of pre- and post-consumer mass by food category, and estimation of embodied greenhouse gas emissions. Renew Agric Food Syst 31(3):191–201

Daly H (1990) Commentary: toward some operational principles of sustainable development. Ecol Econ 2:1–6

Davis J, Mengersen K, Bennet S, Mazerolle L (2014) Viewing systematic reviews and meta-analysis in social research through different lenses. SpringerPlus 3:511

Delley M, Brunner TA (2018) Household food waste quantification: comparison of two methods. Br Food J 120(7):1504–1515

Djekic I, Operta S, Djulancic N, Lorenzo JM, Barba FJ, Djordjevic V, Tomasevic I (2019) Quantities, environmental footprints and beliefs associated with household food waste in Bosnia and Herzegovina. Waste Manag Res 37(12):1250–1260

Ekardt F (2024) Foundations in natural science, economics and epistemology: problems, categories, strategies, and the issue of growth. In: Sustainability. Environmental humanities: Transformation, governance, ethics, law. vol 5. Springer, Cham. https://doi.org/10.1007/978-3-031-62711-8_1

Elimelech E, Ayalon O, Eyal E (2018) What gets measured gets managed: a new method of measuring household food waste. Waste Manag 76:68–81
Elimelech E, Ert E, Ayalon O (2019) Bridging the gap between self-assessment and measured household food waste: a hybrid valuation approach. Waste Manag 95:259–270
Eriksson M, Strid I, Hansson PA (2012) Food losses in six Swedish retail stores: wastage of fruit and vegetables in relation to quantities delivered. Resour Conserv Recycl 68:14–20
Eriksson M, Malefors C, Bergstrom P, Eriksson E, Osowski CP (2020) Quantities and quantification methodologies of food waste in Swedish hospitals. Sustainability 12:3116
Eriksson M, Christensen J, Malefors C (2023) Making food waste illegal in Sweden—potential gains from enforcing best practice in the public catering sector. Sustain Prod Consumption 35:229–237. https://doi.org/10.1016/j.spc.2022.11.003
European Commission (2015) Closing the loop—an EU action plan for the circular economy. Access date 15 June 2024. Available from: https://eur-lex.europa.eu/resource.html?uri=cellar:8a8ef5e8-99a0-11e5-b3b7-01aa75ed71a1.0012.02/DOC_1&format=PDF
Eurostat (2022a) Circular economy. Monitoring framework. Access date 15 June 2024. Available from: https://ec.europa.eu/eurostat/web/circular-economy/monitoring-framework
Eurostat (2022b) Food waste: 127 kg per inhabitant in the EU in 2020. Access date 15 June 2024. Available from: https://ec.europa.eu/eurostat/web/products-eurostat-news/-/ddn-20220925-2
FLW Protocol (2016) Guidance on FLW quantification methods. Access date 15 June 2024. Available from: http://flwprotocol.org/wp-content/uploads/2016/06/FLW_Guidance_Chapter7_Surveys.pdf
Food and Agriculture Organization (2014) Food wastage footprint-full-cost accounting. Final Report. FAO, Rome
Food and Agriculture Organization of the United Nations (2011) Global food losses and food waste-extent, causes and prevention. FAO, Rome, Italy
FUSIONS (2014) Definitional framework for food waste. Full report. SIK—The Swedish Institute for Food and Biotechnology, Göteborg, Sweden
FUSIONS (2016) Food waste definition. Access date 15 June 2024. Available from: https://www.eu-fusions.org/index.php/about-food-waste/280-food-waste-definition
Ganglbauer E, Fitzpatrick G, Guldenpfenning F (2015) Why and what did we throw out? Probing on reflection through the food waste diary. In: Proceedings of the 33rd annual CHI conference on human factors in computing systems, association for computing machinery, pp 1105–1114
Garske B, Heyl K, Ekardt F, Weber LM, Gradzka W (2020) Challenges of food waste governance: an assessment of European legislation on food waste and recommendations for improvement by economic instruments. Land 9:231. https://doi.org/10.3390/land9070231
Giboreau A, Schwartz C, Morizet D, Meiselman HL (2019) Measuring food waste and consumption by children using photography. Nutrients 11(10):2410
Giordano C, Piras S, Boschini M, Falasconi L (2018) Are questionnaires a reliable method to measure food waste? A pilot study on Italian households. Br Food J 120(12):2885–2897
Giordano C, Alboni F, Falasconi L (2019a) Quantities, determinants, and awareness of households' food waste in Italy: a comparison between diary and questionnaires quantities. Sustainability 11(12):3381
Giordano C, Alboni F, Cicatiello C, Falasconi L (2019b) Do discount food products end up in the bin? An investigation into the link between deal-prone shopping behavior and quantities of household food waste. Int J Consum Stud 43(2):199–209
Greenhalgh T, Robert G, Macfarlane F, Bate P, Kyriakidou O (2004) Diffusion of innovations in service organizations: systematic review and recommendations. Milbank Q 82(4):581–629. https://doi.org/10.1111/j.0887-378X.2004.00325.x
Gruen R, Elliott J, Nolan M, Lawton P, Parkhill A, McLaren C, Lavis JN (2008) Sustainability science: an integrated approach for health-programme planning. Lancet 372(9649):1579–1589. https://doi.org/10.1016/S0140-6736(08)61659-1

Gustavsson J, Cederberg C, Sonesson U, Emanuelsson A (2011a) The methodology of the FAO study: "Global food losses and food waste—extent, causes and prevention". FAO. SIK, Swedish Institute for Food and Biotechnology, Göteborg, Sweden
Gustavsson J, Cederberg C, Sonesson U, van Otterdijk R, Meybeck A (2011b) Global food losses and food waste—extent, causes and prevention. Food and Agriculture Organization of the United Nations (FAO), Rome, Italy
Hall KD, Guo J, Dore M, Chow CC (2009) The progressive increase of food waste in America and its environmental impact. PLoS ONE 4:1–11
Hanssen OJ, Syversen F, Sto E (2016) Edible food waste from Norwegian households—detailed food waste composition analysis among households in two different regions in Norway. Resour Conserv Recycl 109:146–154
Hartikainen H, Mogensen L, Svanes E, Franke U (2018) Food waste quantification in primary production—the Nordic countries as a case study. Waste Manag 71:502–511
Ilakovac B, Voca N, Pezo L, Cerjak M (2020) Quantification and determination of household food waste and its relation to sociodemographic characteristics in Croatia. Waste Manag 102:231–240
Johnson LK, Dunning RD, Bloom JD, Gunter CC, Boyette MD, Creamer NG (2018) Estimating on-farm loss at the field level: a methodology and applied case study on a North Carolina farm. Resour Conserv Recycl 137:243–250
Kallbekken S, Saelen H (2013) Nudging hotel guests to reduce food waste as a win-win environmental measure. Econ Lett 119(3):325–327
Kasza G, Dorkò A, Kunszabò A, Szakos D (2020) Quantification of household food waste in Hungary: a replication study using the FUSIONS methodology. Sustainability 12(8):3069
Katajajuuri JM, Silvennoinen K, Hartikainen H, Heikkila L, Reinikainen A (2014) Food waste in the Finnish food chain. J Clean Prod 73:322–329
Khalid S, Naseer A, Sahid M, Shah GM, Ullah MI, Waqar A, Abbas T, Imran M, Rehman F (2019) Assessment of nutritional loss with food waste and factors governing this waste at household level in Pakistan. J Clean Prod 206:1015–1024
Langeley J, Yoxall A, Heppel G, Rodriguex EM, Bradbury S, Lewis R, Luxmoore J, Hodzic A, Rowson J (2010) Food for thought? A UK pilot study testing a methodology for compositional domestic food waste analysis. Waste Manag Res 28(3):220–227
Le HS, Matsuy Y, Do TTT, Nguyen PT (2015) Estimation of waste generation on food and beverage service sectors in Hue City, Vietnam. In: Proceedings of international conference of environmental science and energy engineering (ICESEE 2015). Destech Publications
Leverenz D, Moussawel S, Maurer C, Hafner G, Schneider F, Schmidt T, Kranert M (2019) Quantifying the prevention potential of avoidable food waste in households using a self—reporting approach. Resour Conserv Recycl 150:104417
Malefors C, Callewaert P, Hansson PA, Hartikainen H, Pietiläinen O, Strid I, Strotmann C, Eriksson M (2019) Towards a baseline for food waste quantification in the hospitality sector, quantities and data processing criteria. Sustainability 11(13):3541
Mandasari P (2017) Quantifying and analyzing food waste generated by Indonesian undergraduate students. In: Proceedings of international conference on green agro-industry and bioeconomy (ICGAB 2017)
McColl-Kennedy JR, Snyder H, Elg M, Witell L, Helkkula A, Hogan SJ, Anderson L (2017) The changing role of the health care customer: review, synthesis and research agenda. J Serv Manag 28(1):2–33
Miśniakiewicz M, Amicarelli V, Chrobak G, Górka-Chowaniec A, Bux C (2024) Do living arrangements and eating habits influence university students' food waste perception in Italy and Poland? Sustainability 16:2102
Møller H, Ole JH, Gustavsson J, Oestergren K, Stenmarck A (2014) Report on review of (food) waste reporting methodology and practice. Ostfold Research, Goteborg
Monier V, Mudgal S, Escalon V, O'Connor C, Gibon T, Anderson G, Montoux H, Reisinger H, Dolley P, Ogilvie S (2010) Preparatory study on food waste across EU 27. Final Report. European Commission (DG ENV), Brussels, Belgium

Moore JE, Mascarenhas A, Bain J, Straus SE (2017) Developing a comprehensive definition of sustainability. Implement Sci 12:110. https://doi.org/10.1186/s13012-017-0637-1
Moreno LC, Tran T, Potts MD (2020) Consider a broccoli stalk: how the concept of edibility influences quantification of household food waste. J Environ Manag 256:109977
Official Journal of the European Communities. Council Directive of 15 July 1975 on waste. Access date 15 June 2024. Available from: https://eur-lex.europa.eu/legal-content/EN/TXT/PDF/?uri=CELEX:31975L0442&from=EN
Official Journal of the European Union. Commission Delegated Decision (EU) 2019/1596 of 3 May 2019. Access date 15 June 2024. Available from: https://eur-lex.europa.eu/legal-content/EN/TXT/PDF/?uri=CELEX:32019D1597
Official Journal of the European Union. Directive 2008/98/EC of the European Parliament and of the Council of 19 November 2008 on Waste and Repealing Certain Directives. OJ L 312/3; European Union, Brussels, Belgium
One Planet Network (2024) Target 12.3 food loss & waste. Access date 15 June 2024. Available from: https://sdg12hub.org/sdg-12-hub/see-progress-on-sdg-12-by-target/123-food-loss-waste#indicator-item-12.3.1a
Page MJ, McKenzie JE, Bossuyt PM, Boutron I, Hoffmann TC, Mulrow CD et al (2021) The PRISMA 2020 statement: an updated guideline for reporting systematic reviews. BMJ 372(71):1–9. https://doi.org/10.1136/bmj.n71
Pflugh Prescott M, Herritt C, Bunning M, Cunningham-Sabo L (2019) Resources, barriers, and tradeoffs: a mixed methods analysis of school pre-consumer food waste. J Acad Nutr Diet 119(8):1270–1283
Pirani SI, Arafat HA (2016) Reduction of food waste generation in the hospitality industry. J Clean Prod 132:129–145
Quested TE, Palmer G, Moreno LC, McDermott C, Schumacher K (2020) Comparing diaries and waste compositional analysis for measuring food waste in the home. J Clean Prod 262:121263
Rajan J, Fredeen AL, Booth AL, Watson M (2017) Measuring food waste and creating diversion opportunities at Canada's green university. J Hunger Environ Nutr 13(4):573–586
Richter B, Bokelmann W (2017) Explorative study about the analysis of storing, purchasing and wasting food by using household diaries. Resour Conserv Recycl 125:181–187
Schneider F, Part F, Gobel C, LangenGerhards NC, Kraus GF, Ritter G (2019) A methodological approach for the on-site quantification of food losses in primary production: Austrian and German case studies using the example of potato harvest. Waste Manag 86:106–113
Silvennoinen K, Katajajuuri JM, Hartikainen H, Heikkila L, Reinikainen A (2014) Food waste volume and composition in Finnish households. Br Food J 116(6):1058–1068
Snyder H (2019) Literature review as a research methodology: an overview and guidelines. J Bus Res 104:333–339
Sterner RW, Small GE, Hood JM (2011) The conservation of mass. Nat Educ Knowl 3(10):1–20
Stirman S, Kimberly J, Cook N, Calloway A, Castro F, Charns M (2012) The sustainability of new programs and innovations: a review of the empirical literature and recommendations for future research. Implement Sci 7(17). https://doi.org/10.1186/1748-5908-7-17
Strotmann C, Friedrich S, Kreyenschmidt J, Teitscheid P, Ritter G (2017) Comparing food provided and waster before and after implementing measures against food waste in three healthcare food service facilities. Sustainability 9(8):1409
Szabò-Bòdi B, Kasza G, Szakos D (2018) Assessment of household food waste in Hungary. Br Food J 120(3):625–638
Torraco RJ (2005) Writing integrative literature reviews: guidelines and examples. Hum Resour Dev Rev 4(3):356–367
Tostivint C, Oestergren K, Quested T, Soethoudt H, Stenmarck A, Svanes E, O'Connor C (2016) Food waste quantification manual to monitor food waste amounts and progressions. BIO by Deloitte, Sustainability Services, Deloitte Conseil, Neuilly-sur-Seine, France

United Nations (1987) Report of the World Commission on Environment and Development. Out common future. Access date 12 Nov 2024. Available from: https://www.are.admin.ch/are/en/home/media/publications/sustainable-development/brundtland-report.html
United Nations (2024) Progress towards the sustainable development goals. Report of the Secretary-General. Access date 15 June 2024. Available from: https://sdgs.un.org/goals
United Nations Environment Programme (2024) Think eat save. Tracking progress to halve global food waste. Food Waste Index Report 2024, Nairobi, Kenya
van der Werf P, Seabrook JA, Gilliland JA (2018) The quantity of food waste in the garbage stream of southern Ontario, Canada households. PLoS ONE 13(6):0198470
van Dooren C, Mensink F, Everstein K, Schrijnen M (2020) Development and evaluation of the Eetmaatje cup of rice and pasta as an intervention to reduce food waste. Front Nutr 6:197
van Herpen E, van der Lans IA, Holthuysen N, Nijenhuis-de Vries M, Quested TE (2019) Comparing wasted apples and oranges: an assessment of methods to measure household food waste. Waste Manag 88:71–84
Wang LE, Liu G, Liu XJ, Liu Y, Gao J, Zhou B, Gao S, Cheng SK (2017) The weight of unfinished plate: a survey-based characterization of restaurants food waste in Chinese cities. Waste Manag 66:3–12
Wesana J, Gellynck X, Dora MK, Pearce D, De Steur H (2019) Measuring food and nutritional losses through value stream mapping along the dairy value chain in Uganda. Resour Conserv Recycl 150:104416
Willersinn C, Mack G, Mouron P, Keiser A, Siegrist M (2015) Quantity and quality of food losses along the Swiss potato supply chain: stepwise investigation and the influence of quality standards on losses. Waste Manag 46:120–132

Chapter 2
The Nexus Between Environment, Market, and Legislation in the Foodservice Industry

Abstract Food production, consumption, and waste are responsible for causing negative environmental impacts to the environment. On a global scale, the average carbon footprint of food waste amounts to 3.3 $GtCO_2eq$, which represents approximately 6% of the total greenhouse gas emissions. However, when considering the impacts associated with land use change, deforestation, and organic soil management, its impact increases to 4.4 $GtCO_2eq$. In Italy, the Public Contract Code (Legislative Decree 18 April 2016, n. 50) introduced energy and environmental sustainability criteria in order to achieve the environmental objectives envisaged by the Action Plan for the sustainability in the public sector. Precisely, the Ministry of the Environment adopted technical specifications and contractual clauses, referred to as minimum environmental criteria, for the supply of food commodities in the collective catering sector. The present chapter highlights the interplay between legislation, the market, and the environment through a comprehensive examination of the minimum environmental criteria in Italy. The focus is on: (a) the approach of the minimum environmental criteria for the achievement of environmental objectives; (b) the general guidelines for contracting authorities; and (c) the environmental criteria for awarding collective catering in schools, offices, universities, barracks, and hospital facilities. It demonstrates that the enhancement of the nexus between legislation, market and environment could lead both to direct (e.g., reduction of emissions, minimization of waste) and indirect (i.e., improvement of human well-being) benefits at the global and the local scale.

Keywords Foodservice legislation · Foodservice market · Minimum environmental criteria · Organic food

2.1 Introduction

The current agri-food supply chain must adopt a sustainable (i.e., economic, environmental and social sustainability) approach to reach a fair, environmentally friendly and healthy food system, and the measurement and monitoring of resource and

C. Bux, *Strategies for a Sustainable and Environmentally Friendly Hospital Foodservice Management*, Environmental Humanities: Transformation, Governance, Ethics, Law 6,
https://doi.org/10.1007/978-3-031-82050-2_2

waste streams represents an essential basis toward this achievement. The agri-food chain, from upstream to downstream stages, is responsible for negative environmental impacts such as resource depletion and environmental pollution (Krstić et al. 2024; Luthin et al. 2023), mainly associated with intensive agricultural operations. Latest statistics reveal that about 48 million km^2 (more than half of the habitable land) is addressed to crops production, and that greenhouse gas emissions due to the food systems exceed 52.4 $GtCO_2eq$, distinguishing between those related to food products (13.7 $GtCO_2eq$) and those with non-food products (38.7 $GtCO_2eq$) (Ritchie and Roser 2021; Rana et al. 2023). In Europe, soon after Germany and France, Italy is responsible for 98M $MtCO_2eq$ per year. On the global scale, China highly contributes to the entire pollution, being responsible for 2.4 $GtCO_2eq$ (17% of the global food products) (Ritchie and Roser 2021).

Several studies estimated the environmental impacts associated with food commodities, distinguishing between downstream (i.e., primary production, harvest and post-harvest), core (i.e., industrial transformation, distribution) and downstream (i.e., foodservice, household consumption) stages, also considering food waste disposal and valorization alternatives (e.g., compost, biogas generation). Previous literature on the environmental impacts of food production and consumption focus on entire food baskets or specific food commodities. Considering single food commodities, beef (for meat production) is responsible for the highest quota of emissions, being estimated an amount of 99.48 $kgCO_2eq$ per kg of product, followed by lamb with 39.72 $kgCO_2eq$ and beef (for dairy production) with 33.3 $kgCO_2eq$. On the contrary, plant-based products are responsible to low quantities of carbon emissions, such as maize (1.7 $kgCO_2eq$), wheat and rye (1.57 $kgCO_2eq$), and potatoes (0.46 $kgCO_2eq$) (Poore and Nemecek 2018; Ritchie et al. 2022). Qin and Horvath (2021), according to a cradle-to-grave approach, viz. from farm production to household consumption, measured the environmental impacts of specific commodities such as avocado, celery, lemon and strawberry, with strawberries being the most responsible of greenhouse gas emissions per kg of product (0.13 $kgCO_2eq$). In terms of farm production, Winans et al. (2022) highlighted that food lost on field (i.e., tomatoes, lettuce, peaches) contributes to a significant proportion of greenhouse gases, and that specific efforts to reduce their amount could contribute to the entire sustainability of the agri-food chain. Last, also fast-foods contribute to negative environmental impacts, as outlined by Williams and Wikstrom (2011) for the assessment of the hamburger food waste: (i) 0.79 $kgCO_2eq$ per kg of ketchup; (ii) 0.61 $kgCO_2eq$ per kg of bread; (iii) 8.5 $kgCO_2eq$ per kg of cheese; and (iv) 14 $kgCO_2eq$ per kg. However, packaging could highly contribute to the reduction of emissions.

In the field of foodservice, that considers either business activities (i.e., companies responsible for meals preparation), public institutions or charities, Damiani et al. (2021) highlights that the redistribution of food surplus is basically regulated, but in substance is quite complex and significant efforts to optimize redistribution are required to avoid about 1.87 $kgCO_2eq$ per kg of leftovers. Similar results were estimated by Tonini et al. (2018) that quantified the environmental impacts of avoidable food waste generated by processing, wholesale, retail, and foodservice in the United Kingdom. Compared to their activities, foodservice is responsible for about 12% of the greenhouse emissions of the entire food chain, with an amount of 2100 ktCO_2eq emissions per year. In terms of sustainable strategies, although biogas production or bioconversion represent the most sustainable valorization pathways of food waste (i.e., 66 kgCO_2eq and 71 kgCO_2eq per ton of food waste), followed by composting with 99 kgCO_2eq per ton (Mondello et al. 2017), prevention and minimization strategies of food waste represent the most sustainable solution and are more likely to outcompete all alternative valorization pathways. However, the environmental impacts of food waste vary according to food waste composition (i.e., qualitative characteristics), but still some generalization could be done. First, landfilling is the worst available option for all foods (i.e., landfill emissions are particularly high for energy dense foods), whereas if food is unfitted for donation to charities, conversion to animal feed is the best available option (Mouly et al. 2018). Basically, (i) food waste impacts the environment twice: the first time when it is produced and the second time when it is disposed; and (ii) donation could represent a suitable option for avoiding an excessive (and intensive) food production and save food from being thrown away. Nevertheless, also some disposal options, such as composting, biogas production or bioconversion, could have a positive impact on the environment (Amicarelli et al. 2021). Considering Mouly et al. (2018) that estimated net emissions associated with different disposal pathways in five type of supermarkets in the United Kingdom, donation allows for a reduction of – 5583 kgCO_2eq, which is about + 93% compared to animal feeding, + 94% compared to anaerobic digestion, + 99% compared to landfilling and + 153% compared to landfilling.

By definition, foodservice is defined as "the activity that includes the purchase of food and drinks, the preparation of meals with the foodstuffs purchased, the transport and administration of meals, the cleaning of the canteen, the premises of the cooking center and the equipment and crockery used, as well as the management of food surpluses deriving from the preparation and administration of meals" (GURI 2020). The "minimum environmental criteria for the foodservice and foodstuff" (trad. from the Italian: "Criteri ambientali minimi per il servizio di ristorazione collettiva e fornitura di derrate alimentari", herein CAMs), introduced by the Ministerial Decree no. 65 of 9 May 2020 in Italy, identifies different foodservice categories, viz. schools, offices, universities, barracks, hospitals and welfare, social-health and prison facilities. These guidelines for sustainable strategies, reduction of food waste and minimization of the environmental impacts focus on the foodservice industry since it is either essential for promotion healthy and nutritional food habits, or responsible for increasing economic, social and environmental impacts. Historically, the interest toward sustainable production and consumption patterns, as well as toward food waste minimization and in general toward climate change minimization was raised in 1987, when the concept of sustainable development was defined, as follows: "the development that meets the needs of the present without compromising the ability of future generations to meet their own need" (United Nations 1987). Specifically, as outlined by Daly (1990), sustainable development requires that harvest rates are equal to regeneration rates of natural resources, and that waste emission rates are equal to the natural assimilative capacities of the ecosystem into which the waste streams are released.

In 2021, the market size of the global foodservice industry was about 2.52 trillion USD, with an expected increase up to 4.43 trillion USD by 2028 (PR Newswire 2022). In Europe, the catering sector has an annual turnover of approximately 473 billion USD (18% of the global amount) (Statista 2024a, b), for an amount of over 8.32 million employees in food and beverage activities, with Germany at first place (1.67 million), followed by Spain (1.27 million) and Italy (1.26 million). From the environmental perspective, several negative environmental impacts are associated with the entire lifecycle of the foodservice, from natural resource (e.g., energy, water) consumption during the upstream and core stages until the waste (and food waste) generation during the downstream ones (Meier et al. 2021; Mistretta et al. 2019).

In Italy, the CAMs related to the foodservice were introduced by the Public Contract Code (Legislative Decree 18 April 2016, n. 50) to pursue the environmental strategies envisaged by the Action Plan for the sustainability in the public sector (Law 296/2006) and are based on the lifecycle thinking approach (Jacob-Lopes et al. 2021). In light of tackling climate change, such criteria take care of food commodities production, as well as their distribution, packaging, meals preparation and waste disposal, both under the product and the process side (GURI 2020). As outlined by the reference legislation (Ministero della Transizione Ecologica 2022; Ministero dell'Ambiente e della Tutela del Territorio e del Mare 2024), the CAMs are inspired by the environmental impact assessment of food commodities along their entire lifecycle, considering the market context, the business models adopted by specific organization and the current technologies available in Italy. Specifically, the CAMs aim to: (a) support sustainable agricultural and livestock models, as to not compromise soil fertility, reduce energy consumption and dependence on fossil fuels, as well as tackle water pollution and protect biodiversity and animal welfare; (b) safeguard the biodiversity of fish species; (c) promote a diet with lower consumption of animal proteins; (d) minimize food waste and identify sustainable pathways to valorize food surpluses; (e) prevent the generation of waste associated with the collective catering; (f) reduce energy consumption and related climate-changing gas emissions; and (g) support local economies and small producers.

The CAMs were developed to pursue either the environmental, the economic or the social sustainability goals. Theoretically, a holistic interpretation of the CAMs entails an in-depth analysis of the legislative strategies adopted at the international and national level, namely the Sustainable Development Goals (SDGs) by the United Nations (2023), as well as the Green Deal, the Common Agricultural Policy (CAP) and the National Recovery and Resilience Plan (NRPP) at the European and the Italian level, respectively (European Commission 2023; Prota and Viesti 2022). Hence, the present chapter identifies the pillars included in the CAMs and links them to the legislative strategies at the global and the local scale, by carrying out a content analysis. It represents an original research in the collective catering sector, highlighting the nexus between environment, market and legislation, both direct (e.g., reduction of emissions, minimization of waste) and indirect (i.e., improvement of human well-being).

2.2 Theoretical Background and Literature Review

2.2.1 Legislative Framework on the CAMs in Italy

The CAMs define the environmental requirements for various stages of the supply chains and their aim is to identify the best design solution for either products or services under an environmental perspective according to the lifecycle thinking and the market availability. Such CAMs are based on the Plan for the environmental sustainability in the public sector (so-called National Plan on the Green Public Procurement) and were implemented by the Ministry of Ecological Transition (Ministero dell'Ambiente e della Tutela del Territorio e del Mare 2024). The systematic and homogeneous application of the CAMs on the Italian area allows for the spread of environmental technologies and environmentally friendly products and processes, as to leverage the market and boost sustainable performances among economic operators (Ministero dell'Ambiente e della Sicurezza Energetica 2021).

In Italy, the effectiveness of the CAMs was ensured through the article 18 of the Law 221/2015 and, subsequently, through the article 34 containing "Energy and environmental sustainability criteria" of the Legislative Decree 50/2016 "Procurement Code" (amended by Legislative Decree 56/2017), which made its application mandatory by all contracting stations. Such rules ensure that the national policy on green public procurement is incisive not only in the objective of reducing environmental impacts, but in the objective of promoting more sustainable and circular production and consumption patterns, as well as of spreading green employment. Further, the adoption of the CAMs responds to the need of the public sector to rationalize its consumption and reduce expenditure. To date, the CAMs encompass 18 categories of supplies and assignments, as follows: (i) interior furnishings; (ii) street furniture; (iii) incontinence aids; (iv) work shoes and leather accessories; (v) paper; (v) cartridges; (vii) building sector; (viii) cultural events; (ix) public lighting (supply and design); (x) public lighting (service); (xi) lighting, heating/cooling for buildings; (xii) industrial washing and rental of textiles and mattresses; (xiii) municipal waste and street sweeping; (xiv) collective catering; (xv) disinfection; (xvi) printers, textiles; (xvii) vehicles; (xviii) public green.

2.2.2 Literature Review on the Environmental Sustainability in the Foodservice Industry

Foodservice relies on different economic sectors, namely primary production, product manufacturing, distribution, preparation and administration of meals, and other logistics activities related to waste management. Such activities, across the entire supply chain, generate high amounts of environmental impacts, also due to the use of hobs, ovens, refrigerators, crockery, tablecloths, pots, and freezers for the core operations of foodservice (i.e., preparation, conservation and delivery of meals).

According to the IPCC (2019), the agri-food systems contribute to the worldwide greenhouses emissions up to 37%, among which livestock breeding, rice cultivation and fertilizer use account to a high extent. It results that the emissions associated with agriculture are about 13.7 $GtCO_2eq$ (Ritchie and Roser 2021). With regards to the energy consumption for end use, the agri-food system accounts for 26% of the entire energy use at the European level, whereas Italy consumer approx. 13% (of which 85% are from fossil fuels) IPCC (2019). In the field of the energy consumption for end use, the ENEA (2017) asks organizations to measure and reduce (or rationalize) the energy used for the production, the transformation and the conservation of animal and vegetable-based food products, as well as the energy consumed for the functioning of facilities and for the air conditioning of production and processing environments. Last, one essential aspect is related to the production of waste, specifically food waste, which account to up 40% in schools and up to 30% in hospital caterings, and whose reduction requires suitable strategies (Holt-Giménez 2017).

In the last years, a great number of studies was carried out in the field of environmental sustainability and waste management at the school level. Specifically, the trending topics have been related to food waste reduction through education (Piras et al. 2023; Marques et al. 2022; Antón-Peset et al. 2021) and through public initiatives (Pancino et al. 2021; Tregear et al. 2022), as well as environmental and nutritional benefits associated with low-carbon meals (Batlle-Bayer et al. 2021). For instance, Piras et al. (2023) highlighted the need to monitor pupil behavior and the effects of education interventions on their reported food waste, whereas Antón-Peset et al. (2021) explored the influence of didactic interventions, proposing some suitable opportunities to reduce food waste amounts at mid-morning breaks and lunch. In addition, Marques et al. (2022) also discussed the significant role of nutritionists' education to reduce the amount of vegetable waste among students.

2.3 Methodology

The research is based on official documents published by the Italian Ministry of the Ecological Transition, the Ministry of the Environment and the Protection of the Territory and the Sea, and the Ministry of the Environment and Energy Security over time. Specifically, three documents were at the basis of the analysis, namely the Public Contract Code (Legislative Decree 18 April 2016, n. 50), the Action Plan for the sustainability in the public sector (Law 296/2006) and the accompaniment report to the Ministerial Decree 10 March 2020, G.U. 4 April 2020, n. 90. By applying a thematic analysis, the research focuses on the subsequent key items in the collective catering: (i) organic foods (and organic fruit and vegetables); (ii) low animal protein diets; (iii) reusable tableware; (iv) pre-processed foods; (v) fifth-range products; (vi) single-dose foods; and (vii) local foods (and zero-km foods).

2.4 Results and Discussion

The CAMs identify sustainability guidelines and strategies for two key elements, namely foodservice and for foodstuffs. By definition, as previously stated, foodservice is "the activity that includes the purchase of food and drinks, the preparation of meals with the foodstuffs purchased, the transport and administration of meals, the cleaning of the canteen, the premises of the cooking center and the equipment and crockery used, as well as the management of food surpluses deriving from the preparation and administration of meals", whereas foodstuff intends "fruit and vegetables, fish, milk and dairy products, meat and derivatives, eggs and other processed food products" (GURI 2020). In synthesis, the CAMs adopt a lifecycle approach and aim at: (i) support agricultural and livestock production models that are more sustainable from an environmental perspective, in order not to compromise soil fertility, reduce energy consumption and dependence on fossil fuels, as well as reduce eutrophication and water pollution, and protect biodiversity; (ii) safeguard the biodiversity of fish species, imposing the administration of less over-exploited species caught in the sea and encouraging the use of farmed fish, since fishing is currently defined as "unsustainable"; (iii) promote, especially in the canteens of offices, universities and barracks, a diet with lower consumption of animal proteins, since intensive livestock farming represents one of the main causes of deforestation and absorbs approximately 55% of water resources worldwide; (iv) dedicate strategies to avoid (or recover) food surpluses (and food waste), proposing synergies along the entire agri-food chain and collaborations with strategic actors, such as charities; (v) prevent waste through the use, for example, of reusable cutlery and crockery, or the reduction of single-dose products or non-recyclable packaging; (vi) reduce energy consumption and the related emissions of climate-altering gases, favoring a foodservice service defined as "less industrialized" through reminders to the supply chain or the energy efficiency of new equipment in cooking centers; (vii) verify more carefully the quality (and conformity) of the product services, providing computerized tools for traceability and communication of information (GURI 2020).

Table 2.1 summarizes the main characteristics of the CAMs in the foodservice. It focuses on some specific variables, namely the use of organic food, low animal protein diets, reusable tableware, pre-processed foods, fifth-range products, single-dose foods and local foods.

Table 2.1 Main characteristics of the CAMs in the foodservice

Strategy	S	OUB	HWSP
Organic foods	40%[a]	40%[a]	40%[a]
Organic fruit and vegetables	60%[a]	60%[a]	60%[a]
Low animal protein diets	NO	YES	YES
Reusable tableware	YES	YES	YES
Pre-processed foods	NO	NO	NO
Fifth-range products	NO	NO	NO
Single dose foods	NO	NO	NO
Local foods	YES	YES	YES

Note S = schools; OUB = offices, universities and barracks; HWSP = hospitals, welfare, social-health and prison facilities

[a]Refer to the previous version of the CAMs, namely those implemented in 2011. The current version modified the amounts of required organic foods, being the choice of the minimum quotas of organic food to be administered to adults up to the contracting station, whereas the quota of organic food to be administered to pediatric and adolescent users (0–19 years old) is at least at 50%

2.4.1 CAMs in Schools

In the field of school foodservice (i.e., nursery schools, kindergartens, primary and secondary schools of the first and second grade), meals must consist of one or more portions of fruit, a side dish, a first and/or second course made entirely of organic food (or with quality brands). Fruit, vegetables, legumes and cereals must be at least 50% organic. The beef must be 50% organic by weight, while the pork must be 10% organic or in possession of a voluntary product certification and the poultry meat 20% by weight. In particular, the alternative quality certifications to organic are that of "animal welfare in breeding, transport and slaughter" and "antibiotic-free breeding" in the case of pork, while in the case of chicken meat the certification of "farming rural in freedom" and "rural in the open" are added. These certifications must be formally attached to the transport and purchase documents and on the packaging. As regards fish, supplies must comply with the EU Regulation no. 1967/2006 (Official Journal of the European Union 2006) and not be included in the categories defined as "endangered", "critically endangered", "vulnerable" or "almost threatened" identified by the International Union for Conservation of Nature. In the context of cured meats and cheese, these food commodities must be supplied for 30% organic by weight or, alternatively, with DOP (i.e., "Denominazione di Origine Protetta", trad. in English: "Protected Designation of Origin") or IGP (i.e., "Indicazione Geografica Protetta", trad. in English: "Protected Geographical Indication") quality certifications. The oil must be 40% organic extra virgin olive oil.

From the food surplus management perspective (i.e., food waste minimization), the CAMs highlight the importance of portioning. They foresee that the portioning takes place by administering the right number of pieces through the use of appropriate utensils such as ladles, scoops and slotted spoons of different sizes. Further, portioning should be based on the age of the users. In addition, together with the issue of portioning, the reduction of food waste needs practitioners to pay attention to the right cooking point, specifying that food must not be burnt, oxidized or overcooked, whereas the analysis of the causes of food surpluses must be conducted at least twice a year through questionnaires. Last, the CAMs suggest some strategies to avoid food waste, namely the variation of menus that are not appreciated, in any case maintaining the nutritional requirements established by the regulation, as well as the implementation of projects aimed at promoting the culture of food and the diffusion of sustainable and healthy behaviors. The effectiveness of these strategies will be empirically tested at Chaps. 3 and 4.

Considering the organic and solid waste management pathways (e.g., composting, recycling, recovering, landfilling, etc.), the CAMs suggest, on the basis of the quality and the typology of food surpluses, alternative opportunities: (i) the donation to non-profit organizations of social utility that carry out free distribution of food products; (ii) the collection of surpluses in the canteen for animal feed, kennels or catteries, or in composting systems near the place where the waste is generated. No mention is done to opportunities of landfilling with energy recovery, anaerobic digestion systems or landfilling, although these strategies should be considered in case of impossibility of the previous ones. As regards the solid waste fraction, the CAMs require the use of reusable tableware, namely glass or plastic glasses, as well as ceramic or porcelain tableware and steel cutlery, in order to avoid the production of disposable waste. Therefore, the installation of a dishwasher for washing the furnishings is mandatory. As regards the containers used for storing food at room temperature or in refrigeration, they must be reusable and, in any case, entirely recyclable, made of compostable or biodegradable material in accordance with UNI EN 13432.

In the field of food supplying, the CAMs suggest the use of short supply chains or zero-km products. "Short supply chain" means both direct sales between primary producers or associations of primary producers and cooking centers, and sales by the company that manages the product distribution platform (GURI 2020). By "zero-km products" it is meant a distance between cultivated land (or breeding site) and the cooking center of a maximum of 200 km, reduced to 150 km in some specific cases (i.e., the case of recourse to the Decree Law of 18 December 2017). In the context of small municipalities, this distance is reduced to 70 km, while in the islands (i.e., Sicily and Sardinia), foods must come from the regional territory.

2.4.2 CAMs in Offices, Universities, and Barracks

In the field of offices, universities and barracks, the CAMs suggests the adoption of menus based on the Mediterranean diet, as identified by the UNI/PdR25:2016 on the "Mediterranean diet UNESCO intangible cultural heritage of humanity—Guidelines for promoting a lifestyle and a culture approach for sustainable development" (UNI 2016). Furthermore, such menus must include at least one vegetarian dish, or one containing vegetable proteins, per day, as well as an exclusively vegetarian dish at least once every two weeks. The menus must be prepared on a seasonal basis.

Considering the organic issue, fruit and vegetables must be organic for at least 20% by weight, or 10% certified under the National Integrated Production Quality System (so-called SQNPI certification) or equivalent. Furthermore, fruit and vegetables must not be fifth range and must follow the seasonality criterion. The eggs must be 100% organic. The beef and poultry meat must be organic for at least 20% by weight, or 10% certified by the National Quality System for Animal Welfare (so-called SQNBA certification), while the pork must be organic for at least 5% by weight. Recomposed, pre-fried, or pre-breaded meat cannot be used. Fish products must be from FAO 37 or FAO 27 areas and must comply with the minimum size set out in annex 3 of Regulation (EC) no. 1967/2006 (i.e., "minimum sizes of marine organisms") (Official Journal of the European Union 2006). Also, they must not be classified as "critically endangered", "endangered", "vulnerable", or "near threatened". The cured meats and cheese must be organic for at least 30% by weight or, if organic food is not available, you must have a DOP or IGP certification. The oil must be 40% organic extra virgin olive oil. The water must be mains or micro-filtered, while if the administration of wine is foreseen, it must have the DOC (i.e., "Denomination of Controlled Origin") or DOCG (i.e., "Denomination of Controlled and Guaranteed Origin") labels.

As regards food waste prevention, flexible menu programming is required in order to: (i) use foods that are about to expire or unserved food; (ii) allow half portions; (iii) make family-bags (i.e., "doggy bags") available for users; and (iv) guarantee the right cooking point and an adequate level of quality. Food surpluses, distinguishing between first courses, second courses, side dishes and fungal dishes, must be measured and monitored over time, and the reasons for food waste must be explored using questionnaires every six months. Furthermore, corrective measures must be implemented, such as the variation of the menu in relation to undesirable dishes (as tested in Chaps. 3 and 4). Unserved food must first be donated to non-profit organizations of social utility, and subsequently destined for animal feed, kennels or catteries, or recovered through composting systems in neighboring areas or in wet fraction collection containers. Overall, information about food waste should be sent to the contract executive director every six months.

2.4.3 CAMs in Hospitals, Welfare, Social-Health, and Prison Facilities

Foodservice in hospital and welfare structures is more complex compared to the previous typologies, since menus must be differentiated according to age groups and pathologies, with reference to the following categories: (i) patients in neonatology and pediatrics wards, assisted in neonatal, pediatric and adolescent age (also in social-healthcare and detention facilities); (ii) patients in wards other than pediatrics, as well as non-pediatric patients in care, detention and similar facilities (e.g., rest homes); (iii) personnel working at the facilities. As regards the issue of food surpluses, they must be monitored by hospital wards or by homogeneous groups of users. In collaboration with hospital staff (e.g., dietitians, food science specialists), the menus can be modified in terms of weights, nutritional intake and based on the results of screening on the nutritional status of patients. The relationship between food safety, nutrition, opportunities of healing and food waste (closely connected to the patient malnutrition rate) is of particular importance and will be the subject of the subsequent chapters of the book, specifically Chaps. 4 and 5.

2.4.4 *Strategies for a Sustainable and Efficient Foodservice Supply Chain*

In foodservice, in light of the definition provided by the CAMs (GURI 2020), an essential role is played by the cooking centers, i.e. the place where foodstuffs are controlled, recorded and stored before transformation into meals, and were food items return at the end of the meal as waste. Some suggestions for developing a sustainable, efficient and environmentally friendly cooking center could be defined on the basis of the CAMs, as follows.

One first principle that cooking centers should adopt is that of "forward movement without crossings", so as to guarantee optimization of the workflow in terms of non-stressful working conditions for the kitchen staff, efficiency in the service and hygiene (i.e., food safety). Specifically, cooking centers should be composed of different rooms/areas distinguishing between specific moments of the chain, namely: (i) receipt; (ii) control; (iii) storage; (iv) preparation; (v) washing; (vi) cooking; (vii) packaging; and (v) transport, according to the Hazard Analysis and Critical Control Point (HACCP) guidelines (U.S. Food and Drug Administration 2024).

In terms of raw ingredients supplying, companies operating in the foodservice should base their menus on local (and zero-km) products, so as to avoid environmental negative consequences and enhance local markets (i.e., valorization of the territory, respect for traditions). Once foodstuffs arrive to the cooking center, several technical steps should be conducted. First, foods should be checked in the reception area (i.e., acceptance area), and the receptionist checks regard the hygienic conditions of the means of transport, as follows: (i) check of any promiscuity or cross-contaminations of foods; (ii) check of the weight of the unloaded foodstuff; (iii) check of the integrity, with a specific report of any holes, bulges of foreign bodies, for instance; (iv) check of the conformity of the labeling; (v) check of the expiring date; (vi) check of the temperature, on a sample basis, of perishable foods through a probe thermometer or an infrared thermometer. In case of non-conformity of the foodstuff, foods must be returned to the supplier and not accepted by the receptionist. Soon after reception and control of foods, storage requires that foodstuffs do not intersect with finished products ready for consumption, so as to guaranteed food safety, and if possible, the secondary packaging (e.g., cardboard, wood) is removed. In general, foods are stored at room temperature or at a controlled temperature.

Core activities in the cooking center begin with the withdrawal of raw ingredients by the operators, which is done according to the meals planned for the day and in accordance with some guidelines, namely: (i) quantities are taken proportionally to the production capacity of the equipment; (ii) time exposure to uncontrolled temperature is limited; (iii) an additional visual check of packaging, labels and organoleptic characteristics is done to foods before transformation. In terms of meal preparation, it takes place in physically and functionally separate areas that are equipped with specific boards, cutting tools, slicers, etc. according to different foods, and kitchen staff must wear gloves and masks, so as to guarantee hygiene. Generally, processing activities take place with doors and windows closed. Cooking activities are designed to preserve the nutritional value and the organoleptic characteristics of foods, and to guarantee healthiness, and transportation of meals to canteens could be carried out according to different methods, viz. "cook-and-chill" and "cook-and-hold" (see Chap. 3). Overarching purpose of the cooking activities is to improve sensorial quality of meals and encourage healthy and environmentally friendly choices among consumers, by balancing both environmental protection and nutritional values, as well as consumer satisfaction (see Chap. 4).

Some best practices could be considered when cooking food, as follows: (i) adoption of healthy cooking techniques, with. a minimal use or absence of fats to prevent hypercholesterolemia; (ii) use of aromatic herbs to flavor meals and to reduce the use of salt; (iii) use of cold extra virgin oil at the end of cooking; (iv) avoidance, if possible, of saturated fats and degreasing of meat; (v) respect of the seasonality, territoriality and traditions of products; (vi) incentive for the consumption of fresh seasonal fruit, vegetables and legumes; (vii) rotation of breads, such as white, with cereals, without salt, etc.

In terms of food transportation to canteens (i.e., logistics), different methods could be considered, as follows: (i) cook-and-chill, which is based on preservation through refrigeration of meals; and (ii) cook-and-hold, in which meals cannot be stored for more than three hours on hot counters and thermal trolleys.

2.5 Conclusions

The introduction of the CAMs in the Italian market has significant scopes, which are essential in boosting economic, social and environmental sustainability, as well as competitiveness among entrepreneurs and public authorities. Specifically, the CAMs: (i) highlight the environmental and social impacts along the life cycle of the collective catering service; (ii) describe the characteristics of the catering service in the market context; (iii) describe possible strategies to reduce the environmental footprint of the catering service either in schools, offices, universities, barracks, hospitals, welfare, social-health, or prison facilities; and (iv) offers indications and technical-methodological clarifications for public authorities and private economic operators. It represents a dynamic document, since several characteristics and possible strategies should be updated in time.

From the social perspective, the introduction of the CAMs represent an essential tool to boost employment and increase well-being. Specifically, their application takes into consideration the condition of workers, especially seasonal workers in farms, as to avoid their exploitation and the unemployment. Further, the CAMs offer indirect support to local economic and small producers by boosting the supplying of zero-km, fair trade and short-chain products, as well as a fair profit margin for both catering companies and farmers. As regards populations characterized by poverty and food insecurity, the CAMs guarantee that they are not deprived from precious food resources in an equitable perspective. Last, the CAMs support the employment of disadvantages or disabled people, as to assure them, although indirectly, opportunities for social integration and well-being.

References

Amicarelli V, Lagioia G, Bux C (2021) Global warming potential of food waste through the life cycle assessment: an analytical review. Environ Impact Assess Rev 91:106677

Antón-Peset A, Fernandez-Zamudio M-A, Pina T (2021) Promoting food waste reduction at primary schools. A case study. Sustainability 13:600

Batlle-Bayer L, Bala A, Aldaco R, Vidal-Monés B, Colomé R., Fullana-i-Palmer P (2021) An explorative assessment of environmental and nutritional benefits of introducing low-carbon meals to Barcelona schools. Sci Total Environ 756:143879

Daly H (1990) Commentary: toward some operational principles of sustainable development. Ecol Econ 2:1–6

Damiani M, Pastorello T, Carlesso A, Tesser S, Semenzin E (2021) Quantifying environmental implications of surplus food redistribution to reduce food waste. J Clean Prod 289:125813

ENEA (2017) Rapporto annuale sull'efficienza energetica 2017. Access date 17 June 2024. Available from: https://www.efficienzaenergetica.enea.it/component/jdownloads/?task=download.send&id=126&catid=40&Itemid=101

European Commission (2023) The common agricultural policy at a glance. Access date 17 June 2024. Available from: https://agriculture.ec.europa.eu/common-agricultural-policy/cap-overview/cap-glance_en. Accessed 17 Apr 2023

GURI (Gazzetta Ufficiale della Repubblica Italiana). Servizio di ristorazione e fornitura di derrate alimentare (approvato con DM n. 65 del 10 marzo 2020, in G.U. n. 90 del 4 aprile 2020). Access date 17 June 2024. Available from: https://www.mase.gov.it/sites/default/files/archivio/allegati/GPP/2021/cam_ristorazione.pdf

Holt-Giménez E (2017) A foodie's guide to capitalism. Understanding the political economy of what we eat. Publisher/Imprint. Monthly Review Press, U.S.

IPCC (2019) Special report on climate change and land. Access date 17 June 2024. Available from: https://www.ipcc.ch/srccl/

Jacob-Lopes E, Queiroz Zepka L, Costa Deprá M (2021) Assistant's tools toward life cycle assessment. In: Jacob-Lopes E, Zepka LQ, Deprá MC (eds) Sustainability metrics and indicators of environmental impact, chap 5. Elsevier, pp 77–90

Krstić M, Elia V, Agnusdei GP, De Leo F, Tadić S, Miglietta PP (2024) Evaluation of the agri-food supply chain risks: the circular economy context. Br Food J 126(1):113–133. https://doi.org/10.1108/BFJ-12-2022-1116

Luthin A, Traverso M, Crawford RH (2023) Circular life cycle sustainability assessment: an integrated framework. J Ind Ecol 28:41–58

Marques C, Lima JPM, Fialho S, Pinto E, Baltazar AL (2022) Impact of a food education session on vegetables plate waste in a Portuguese school canteen. Sustainability 14:16674

Meier T, von Borstel T, Welte B, Hogan B, Finn SM, Bonaventura M, Friedrich S, Weber K, Dräger de Teran T (2021) Food waste in healthcare, business and hospitality catering: composition, environmental impacts and reduction potential on company and national levels. Sustainability 13:3288

Ministero dell'Ambiente e della Sicurezza Energetica (2021) CAM—Criteri ambientali minimi. Access date 17 June 2024. Available from: https://gpp.mite.gov.it/Home/Cam

Ministero dell'Ambiente e della Tutela del Territorio e del Mare. Piano d'azione per la sostenibilità ambientale dei consumi nel settore della pubblica amministrazione. Access date 17 June 2024. Available from: https://www.mase.gov.it/sites/default/files/archivio/allegati/GPP/PAN_GPP.pdf

Ministero della Transizione Ecologica Relazione Accompagnamento (2022) CAM Servizio di ristorazione collettiva e fornitura derrate alimentari (DM 10 marzo 2020). Access date 17 June 2024. Available from: https://gpp.mite.gov.it/PDF/CAM%20Ristorazione%20Relazione%20accompagnamento%20aprile%202022.pdf

Mistretta M, Caputo P, Cellura M, Cusenza MA (2019) Energy and environmental life cycle assessment of an institutional catering service: an Italian case study. Sci Total Environ 657:1150–1160

Mondello G, Salomone R, Ioppolo G, Saija G, Sparacia S, Lucchetti MC (2017) Comparative LCA of alternative scenarios for waste treatment: the case of food waste production by the mass-retail sector. Sustainability 9(5):827

Mouly JA, Allan SR, Hewitt CN, Berners-Lee M (2018) Greenhouse gas emissions of food waste disposal options for UK retailers. Food Policy 77:50–58

Official Journal of the European Union (2006) Council Regulation (EC) No. 1967/2006 of 21 December 2006 concerning management measures for the sustainable exploitation of fishery resources in the Mediterranean Sea. Access date 17 June 2024. Available from: https://eur-lex.europa.eu/legal-content/EN/TXT/PDF/?uri=CELEX:32006R1967

Pancino B, Cicatiello C, Falasconi L, Boschini M (2021) School canteens and the food waste challenge: which public initiatives can help? Waste Manag Res 39(8):1090–1100

Piras S, Righi S, Banchelli F, Giordano C, Setti M (2023) Food waste between environmental education, peers, and family influence. Insights from primary school students in Northern Italy. J Cleaner Prod 383:135461

Poore J, Nemecek T (2018) Reducing food's environmental impacts through producers and consumers. Science 360(6392):987–992

PR Newswire (2022) Market size of the global food service industry in 2021, with a forecast for 2028 (in trillion U.S. dollars). Access date 17 June 2024. Available from: https://www.statista.com/statistics/1095667/global-food-service-market-size/

Prota F, Viesti G (2022) Linking the 'recovery and resilience plan' and smart specialisation. The Italian case. JRC working papers on territorial modelling and analysis, European Commission, Seville, JRC130071

Qin YW, Horvath A (2021) Contribution of food loss to greenhouse gas assessment of high-value agricultural produce: California production US consumption. Enrivon Res Lett 16:014024

Rana RL, Bux C, Lombardi M (2023) Carbon footprint of the globe artichoke supply chain in Italy: from agricultural production to industrial processing. J Clean Prod 391:136240

Ritchie H, Rosado P, Roser M (2022) Environmental impacts of food production. Access date 17 June 2024. Available from: https://ourworldindata.org/environmental-impacts-of-food

Ritchie H, Roser M (2021) Environmental impacts of food production. Access date 17 June 2024. Available from: https://ourworldindata.org/environmental-impacts-of-food#citation

Statista (2024a) Net turnover of the food and beverage service activities industry in the European Union (EU-27) in 2021 and 2022 (in million euros). Access date 17 June 2024. Available from: https://www.statista.com/statistics/1459198/net-turnover-food-beverage-service-activities-industry-eu/

Statista (2024b) Number of employees in the food and beverage service activities industry in Europe in 2022, by country. Access date 17 June 2024. Available from: https://www.statista.com/statistics/1459209/number-of-employees-food-beverage-services-europe-by-country/

Tonini D, Albizzati PF, Astrup TF (2018) Environmental impacts of food waste: learnings and challenges from a case study on UK. Waste Manag 76:744–766

Tregear A, Aničić Z, Arfini F, Biasini B, Bituh M, Bojović R, Brečić R, Brennan M, ColićBarić I, Del Rio D, Donati M, Filipović J, Giopp F, Ilić A, Lanza G, Mattas K, Quarrie S, Rosi A, Sayed M, Scazzina F, Tsakiridou E (2022) Routes to sustainability in public food procurement: an investigation of different models in primary school catering. J Clean Prod 338:130604

U.S. Food and Drug Administration. Hazard analysis critical control point (HACCP). Access date 17 June 2024. Available from: https://www.fda.gov/food/guidance-regulation-food-and-dietary-supplements/hazard-analysis-critical-control-point-haccp

UNI (2016) Dieta mediterranea patrimonio immateriale UNESCO—linee guida per la promozione di uno stile di vita e di una cultura favorevole allo sviluppo sostenibile. Access date 17 June 2024. Available from: https://www.obesityday.org/usr_files/biblioteca/UNESCO_dieta-mediterranea-lineeguida.pdf

United Nations (1987) Report of the World Commission on Environment and Development. Out common future. Access date 17 June 2024. Available from: https://www.are.admin.ch/are/en/home/media/publications/sustainable-development/brundtland-report.html

United Nations (2023) The 17 goals. Access date 17 June 2024. Available from: https://sdgs.un.org/goals

Williams H, Wikstrom F (2011) Environmental impact of packaging and food losses in a life cycle perspective: a comparative analysis of five food items. J Clean Prod 19(1):43–48

Winans K, Marvinney E, Gillman A, Spang E (2022) An evaluation of on-farm food loss accounting in life-cycle assessment (LCA) of four California specialty crops. Front Sustain Food Syst 4:1–10

Chapter 3
Energy Efficiency in the Foodservice Industry

Abstract Healthcare is responsible for curative, preventive, rehabilitative, and palliative care and is one of the fastest-growing industries on the global scale. It is one of the fastest-growing industries on the global scale. However, despite its essential role in society, it has significant negative impacts on the environment, mainly due to energy consumption, transportation, production use and disposal, and waste generation. In this field, best practices in foodservice play an essential role in achieving different aims, such as increasing the energy efficiency of cooking centers and logistics, increasing customer satisfaction, and reducing malnutrition. The current chapter evaluates the energy consumption of a cooking center by comparing cook-and-hold and cook-and-chill service, and then it quantifies the amount of food waste generated in nine different hospitals by measuring consumer satisfaction through a questionnaire. It is possible to reduce diesel and electricity consumption by 42% and 93%, respectively, by adopting the cook-and-hold method compared to the cook-and-chill method. Furthermore, this logistical approach has the potential to enhance customer satisfaction and healing prospects by reducing food waste (and malnutrition) among patients. Cooking centers, kitchen staff, and sustainable diets play an essential role toward environmental, economic, and social sustainability. This chapter presents some strategies for improving foodservice performance through a real and concrete case study.

Keywords Cook-and-chill method · Cook-and-hold method · Customer satisfaction analysis · Diesel consumption · Electricity consumption · Energy efficiency · Food waste analysis

3.1 Introduction

Material flow analysis for measuring inputs and outputs within a specific system, such as food commodities and food waste in foodservice activities, is considered as a "complementary method" for data collection by the latest ISO 59020:2024 on "Circular

C. Bux, *Strategies for a Sustainable and Environmentally Friendly Hospital Foodservice Management*, Environmental Humanities: Transformation, Governance, Ethics, Law 6,
https://doi.org/10.1007/978-3-031-82050-2_3

economy—Measuring and assessing circularity performance" (ISO 2024b). Specifically, as suggested by the standard, the material flow analysis (also defined "resource flow analysis") helps organization quantify resource or material flows (and stocks) between processes, distinguishing between electricity or water streams, as well as commodities ones. However, one overarching challenge relates to data collection (i.e., how data can be sourced), their nature and their origin, and the transparent selection of spatiotemporal boundaries for the analysis. Uncertainties, incomplete inventories, and a lack of expertise restrict the application of the material flow analysis at the organizational level, hindering its widespread adoption.

Material flows measurement plays a key role in reducing food waste at foodservice and households and increasing sustainability along the agri-food supply chain, but despite the growing body of research on sustainable development (and circular economy) in the healthcare sector (Berniak-Woźny and Rataj 2023; Xu et al. 2021), there is still room for research on hospital food consumption and waste patterns. Practically, the measurement and monitoring operations, within the boundaries of a single organization (i.e., micro-level) or a group of them (i.e., meso-level), is required by national and international authorities and should be done in accordance with specific standards, such as the UNI/TS 11820:2022 on "Measurement of Circularity—Methods and Indicators for Measuring Circular Processes in Organizations" (UNI 2022) or the ISO 59020:2024 (ISO 2024b).

The healthcare industry is responsible for energy consumption and pollution to high extents, affecting either human health or the environment (Sherman et al. 2019) and accounting for approximately 5% of the carbon emissions worldwide (NHS 2020). Latest statistics on greenhouse gas emissions in the healthcare sector reveal that their highest quota is associated with the generation and distribution of electricity, gas and heat or cooling activities (40%), followed by healthcare facilities operational emissions (13%), other manufacturing such as basic metals, textiles and food products (11%), agriculture such as provision of food products (9%), other sectors and services (8%), transport for either travels or logistics (7%), pharmaceutical and chemical products (5%), waste treatment (3%) and other residual activities such as plastic products or electronic equipment (Karliner et al. 2024). Worldwide, the United States (547 $MtCO_2eq$), China (342 $MtCO_2eq$) and the European Union (249 $MtCO_2eq$) represent the top three contributors to the healthcare climate footprint (Karliner et al. 2024) but statistics vary if the "per capita emission" is considered. In the United States, the emissions per capita amount to about 1.72 tCO_2eq, whereas in China they descent to 0.25 tCO_2eq and in the European Union the emissions are on average 0.49 tCO_2eq.

The achievement of the sustainability in the healthcare sector must consider that energy, above all energy generated from fossil fuels and non-renewable sources, is responsible for about half of the entire carbon emissions of the sector. But emissions are associated also with waste management, and food waste management as well. In Europe, food waste per capita amounts to about 131 kg per year (Eurostat 2022), but food waste in hospitals exceeds that of any other foodservice sector by a factor 2 or 3 (Dias-Ferreira et al. 2015; Arriz-Jorquiera et al. 2024). On average, hospital food waste amounts to 0.08 kg per patient per day at breakfast, 0.22 kg at lunch and about 0.23 kg at dinner (Paiva et al. 2022), but these statistics are highly variable due to small sample sizes and uncertain measurement method. For instance, Sahal Alharbi et al. (2020) estimated similar quantities for breakfast, but lunch and dinner are highly reduced by about 30%. To this extent, Bouley et al. (2017) identified some actions to reduce the healthcare climate footprint, such as: (i) use of appropriate low-carbon technologies for care; (ii) design or construction of low-carbon or zero emissions buildings; (iii) investment in renewable energy and adoption of energy efficiency strategies; (iv) sustainable waste, water and transport management.

In light of these premises and considering the definition of foodservice presented in the CAMs (i.e., "the activity that includes the purchase of food and drinks, the preparation of meals with the foodstuffs purchased, the transport and administration of meals, the cleaning of the canteen, the premises of the cooking center and the equipment and crockery used, as well as the management of food surpluses deriving from the preparation and administration of meals") (GURI 2020), the present chapter: (i) evaluates the energy consumption of a cooking center by comparing cook-and-hold and cook-and-chill service; and (ii) quantifies the amount of food waste generated in nine different hospitals by measuring consumer satisfaction through a questionnaire, on a sample of 984 patients. The purpose is to identify sustainability strategies for reducing energy consumption, enhance energy efficiency and increase customer satisfaction, that represents an overarching purpose in hospital facilities being food waste, malnutrition and healing opportunities strongly correlated. From the methodological perspective, the chapter illustrates the main features of the energy efficiency analysis, the customer satisfaction analysis and the food waste analysis, proposing both in theory and in practice a mixed method that could be easily applied in other realities (for either practitioners or researchers).

3.2 In-Depth Literature Review on Food Waste in Foodservice

Based on the guidelines provided by Snyder (2019) to carry out systematic and critical reviews and being the PRISMA model used with success in several authoritative articles (Cook et al. 2022; Page et al. 2021), a brief review in the field of food waste in foodservice was conducted. Distinguishing between schools, universities, workplaces and hospitals, the review searched for original research articles in Scopus according to the subsequent criteria: (i) considering the Commission Delegated Decision 1597/2019, the research timeframe was 2019–2023; (ii) considering the need to explore the environmental, social, and economic sustainability related foodservice, the selected subject areas were environmental science, social science, energy, business management, and accounting; and (iii) considering the need to explore dependable insights, only articles published in international journals in English were explored. According to a TITLE-ABS-KEY query, 351 articles emerged by searching for the subsequent research strings: "food waste" AND "school" OR "university" OR

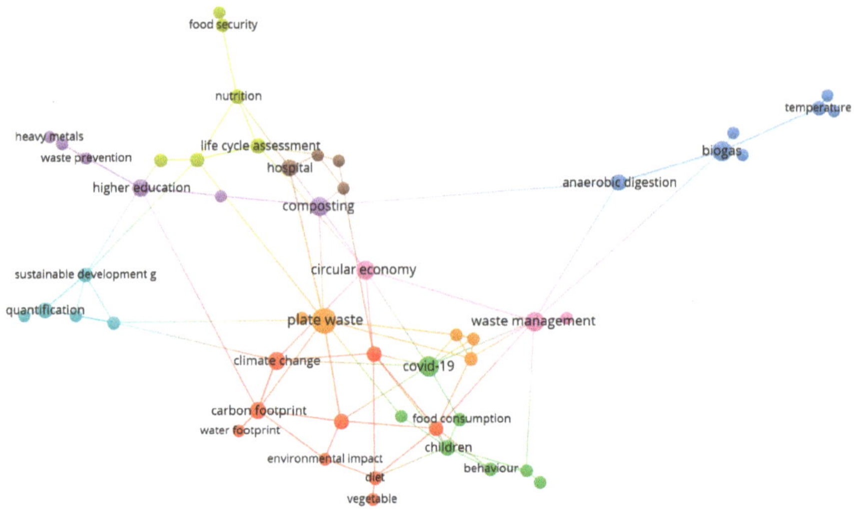

Fig. 3.1 VOSviewer results in the field of food waste in foodservice

"workplace" OR "hospital". Figure 3.1 illustrates the main research trends related to food waste in canteens and catering services, as resulting by the application of the VOSviewer software, which is a suitable tool that provides insights into the structure of the research network and offer a multidimensional scaling and clustering in the research (van Eck and Waltman 2021).

Some research trends were in-depth explored in the field of food waste from the methodological (e.g., life cycle assessment, carbon footprint, water footprint), the consumption behavior, the valorization (e.g., anaerobic digestion, biogas) or the food safety and security perspective. As a common thread, the main recurring words are related to "plate waste", "circular economy", "nutrition", "waste management" and "climate change". However, the quest related to energy efficiency seem rather neglected.

From the bibliographic perspective, the vast majority of articles was published in 2021 (101 contributions), followed by 2022 (94 contributions) and 2020 (77 contributions), whereas in the first trimester of 2023 only 16 articles were published. In the field of publishing journals, the largest number of articles was published in Sustainability (47 contributions) and in the International Journal of Environmental Research and Public Health (21 contributions) by MDPI, whereas several articles were published by Elsevier, as follows: (i) Journal of Cleaner Production (18 contributions); (ii) Resources Conservation and Recycling (17 contributions); (iii) Science of the Total Environment (16 contributions); (iv) Journal of Environmental Management (9 contributions); and (v) Waste Management (8 contributions). From the geographical perspective, the large number of articles regards the analysis of food waste in the United States (63 contributions), followed by China and the United Kingdom (36 contributions each). Italy is at fourth place, accounting for 31 research articles.

In the last years, a plethora of studies was conducted at the school level, with specific reference to food waste reduction through didactic interventions and education (Piras et al. 2023; Marques et al. 2022), food waste environmental and nutritional benefits by introducing low-carbon meals (Batlle-Bayer et al. 2021) and the adoption of public initiatives toward food waste minimization (Pancino et al. 2021; Tregear et al. 2022). One common thread toward food waste minimization at schools is represented by didactic interventions and education. Piras et al. (2023) discussed that theoretical lessons reduce the self-declared food waste in the short-term, but it does not significantly reduce food waste amount in the long-term, whereas imitations through direct observation could play an important role in tackling end-users' food waste. Further, Antón-Peset et al. (2021) investigated the influence of didactic interventions, which could lead to the reduction in the amount of food waste during mid-morning breaks and lunch. Such a research confirmed the importance of educate pupils under the environmental and social perspective, and it can have positive effects on children's attitude in food waste minimization. Last, Marques et al. (2022) estimated the impact of nutritionist's food education sessions on vegetable waste and consumption, highlighting their efficacy toward the reduction of food vegetable leftovers and the increase in per capita vegetable consumption.

In the field of low-carbon meals, Antòn-Peset et al. (2021) estimated that the shift from conventional meals to plant-base and low-carbon diets can mitigate climate change from 46 to 60% in terms of blue water footprint, primary energy demand, land use and global warming potential. Further, the research highlights the need to prevent food waste in canteens, since its emissions amount to approx. 21% of the entire meals' emissions. Among strategies to reduce food waste among pupils, some authors discussed the need to invest of the quality of the meals, as well as to explore some variables such are satiation, meal size, children food choices, location of the kitchen, and social interaction (Boschini et al. 2020; Zhao et al. 2019). Tregear et al. (2022) discussed that the reduction of ruminant meat in menus represents the best measure to reduce greenhouse gases emissions, whereas under the economic perspective one possible measure is to invest in local catering staff. Last, supervision is an important action to reduce food waste among pupils. Further, Pancino et al. (2021) measured food waste quantities in 78 primary school, estimating that over 28% is prepared but not consumed. Under the public authorities' perspective, the research still highlights the need to implement donation programs for non-served food or the use of doggy bags to avoid plate waste disposal, stressing the need to introducing preventive and monitoring actions.

Soon after schools, research focuses on food waste at universities, which still represents a topical concern among university students (Knezevic et al. 2019). On the one side, several authors identified possible strategies to reduce food waste through consumers' behavior (Malefors et al. 2022; Qian et al. 2021), whereas other researchers investigated possible (and sustainable) valorization pathways of food waste (Malefors et al. 2022; Qian et al. 2021). Further, some scholars were interested in the environmental impacts of food waste (Qian et al. 2022; Li et al. 2021).

Considering drivers and determinants of food waste in university foodservice, Qian et al. (2021) highlighted the multidimensional factors, which affect food waste. Among others, the research illustrates the main drivers of food waste, such as gender, education, daily waste habits, family wealth, presence of others, time pressure, food taste satisfaction and location, and provides some opportunities toward its minimization, such as practical and effective restraint, guidance and incentive measures. Further, Malefors et al. (2022) introduced practical interventions to reduce food waste, such as tasting spoons, plate waste trackers and guest forecasting tools. It has been estimated that such interventions could reduce food waste from 6 to 44 g per guest. In the field of sustainable pathways for food waste valorization, one common target in universities is the achievement of decarbonization. Wang et al. (2023) estimated the biomethane potential of food waste in over 2000 university campuses in China, assessing that over 195 million m^3 of biomethane could be obtained from approx. 1.74 million tons of food waste from university canteens through anaerobic digestion. Considering the environmental impacts of biogas production from food waste, Zhou et al. (2021) carried out a life cycle assessment on anaerobic digester and pyrolysis plants, and similar studies were conducted by Qian et al. (2022), which assessed the carbon footprint associated with food waste. It was estimated a carbon footprint of 233 g CO_2eq per capita per day, with meat waste having the highest footprint (approx. 107 g CO_2eq per capita per day), followed by grains and vegetables. Last, Li et al. (2021) evaluated both the environmental and the economic costs associated with food waste in university canteens. Through the life cycle costing method, the research has estimated that food waste costs approx. 330 USD per student per year.

Food waste in workplace (or worksite) canteens was scarcely investigated in the field of the environmental sciences, specifically with regards to users' behavior. Pires et al. (2022) compared food waste perception among workers and food waste amount in plate waste, as to comprehend if informed (and trained) consumers can estimate their food waste. The research highlights the positive relation between self-assessment and measured plate waste among trained consumers, as well as the excessive portions as the main driver for food waste in workplace canteens. Further, Wang et al. (2022) highlighted the need to implement macro-level workplace campaigns to decrease food waste in staff cafeterias, by combining smart bins with fortnightly information feedback. It results that workers receiving an environmental framing are more likely to reduce food waste both at work and at home. Similar studies were conducted by Chawla et al. (2022) and Lorenz-Walther and Langen (2020). Chawla et al. (2022) investigated food waste drivers among employees working in luxury hotels by using the theory of the planned behavior based on personal observation, organizational documents and semi-structured interviews. It appears that workers' pro-environmental behaviors are positively influenced by environmental corporate policies, embedded work practices (e.g., economic value of food waste) and operational pressures (e.g., prices for efficient performances). Last, Lorenz-Walther and Langen (2020) discussed workers' acceptance for sustainable restructuring in their workplace canteens, specifically with reference to personal food choices, lunchtime habits, as well as pre-ordering, payment system and portion sized. It results that out-of-home food habits, compared to household habits, are more influenced by context-factors (e.g., social context, social eating norms, payment methods) and intervention are easier.

Food waste in hospitals represents a scarcely investigated topic in research, and Chap. 5 systematically examines all the studies addressed to the issue.

3.2.1 Energy Efficiency in Foodservice

Basically, the concept of energy efficiency relates to a minor consumption of energy to perform the same task, so it refers to the reduction in the so-called "energy waste". Such as in the field of solid or food waste minimization, also energy efficiency allows for the reduction in greenhouse gas emissions, as well as in the reduction of energy imports and increased economic benefits for companies and households (Environmental and Energy Study Institute 2024). In terms of international standards, the ISO 5000 on "Energy management" (ISO 2024a) helps organizations address their impact, conserve resources and improve the bottom line through an efficiency energy management, and it is strictly related to the ISO 9001 and the ISO 14001 from a holistic perspective of quality, environmental and energy management. Practically, the ISO 50001 offers an instrument for organization to: (i) develop a more efficiency energy policy; (ii) fix target and objectives toward the achievement of such policy; (iii) interpret data for a more efficiency decision-making about energy; (iv) measure the results; and (v) after review, continually improve the energy management (in line with the iterative design "plan-do-check-act").

Historically, the first European standard on energy efficiency was the EN 16001:2009, whose aim was to advance energy efficiency for products, whereas the subsequent ISO 50001:2011 and all its revisions aimed at improving the energy performance, intending efficiency, use and consumption. On the global scale, in 2022, about 27,765 organizations are certified with the ISO 50001, and Italy ranks at fourth place with 1656 organizations (6%) (Accredia 2022).

By definition, the standard identify energy as "electricity, fuel, steam, heat, compressed air and other similar means", whereas energy efficiency is defined as the "ratio or other quantitative relationship between results in terms of performance, services, goods or energy, and energy input" (ISO 2024a). Practically, organization that intend to certificate should also identify the energy use within the boundary of their organization, define the significance (quantitatively) of their energy uses, define the Energy Performance Indicators (EnPI), and catalogue the opportunities for improvement with reference to the various energy aspects. In general, considering a cooking center operating in the foodservice industry, the main energy carriers and the prevalent energy uses relate to: (i) electricity for electric drives, industrial washing (i.e., dishwasher), refrigeration, air conditioning, cold air treatment, lighting; (ii) natural gas for cooking food, domestic hot water and warm air conditioning; and (iii) diesel for transportation of meals from the cooking center to the various facilities.

3.3 Materials and Methods

3.3.1 Research Framework, Goal, and Scope

The present chapter reports the results from a case-study conducted in the Azienda Sanitaria Locale (ASL BA), which is served by four different cooking centers and is composed of nine hospitals and 60 hospital wards located in Altamura, Corato, Bari, Molfetta, Monopoli, Putignano, Terlizzi and Triggiano. It considers the definition of foodservice at outlined by the CAMs (GURI 2020). Specifically, the research focuses on these variables: (i) transportation of meals; (ii) administration of meals; and (iii) management of food surpluses after food consumption among hospital patients. The goals of the research are: (i) evaluate the energy consumption of a cooking center by comparing cook-and-hold and cook-and-chill service; and (ii) quantify the amount of food waste generated in nine different hospitals by measuring consumer satisfaction through a questionnaire, with the overarching purpose of defining sustainable strategies for an increased energy efficiency and customer satisfaction, which translated into a reduction of food waste and malnutrition among patients.

3.3.2 Data Collection

3.3.2.1 Energy Consumption Analysis

The research considers a gate-to-consumer approach, including all energy consumption activities from meal distribution (soon after cooking) to final consumption among consumers. Specifically, it compares two different foodservice methods, namely cook-and-chill and cook-and-hold, so as to identify strengths and weaknesses from the energetic perspective. Figure 3.2 illustrates the system boundaries of the research, focusing on different steps. In terms of cook-and-chill, it considers: (i) blast freezing at 10 °C; (ii) storage in cold rooms; (iii) energy consumption to keep the cold temperature during packaging; (iv) transportation with cook-and-chill trucks; and (v) use of warm trolleys to boost temperature up to 120 °C for 50 min. As regards, cook-and-hold, it takes care of warm storage and transportation with non-refrigerated trucks, as well as the use of warm trolleys.

Primary data for the energy consumption analysis were retrieved from one cooking center, which responds to the legislative requirements foreseen by the Regulation (EC) 853/2004 of 29 April 2004 laying down specific hygiene rules for the hygiene of foodstuff (Official Journal of the European Union 2004) and by the Regional Regulation 2108/2020 that identified "guidelines for hospital foodservice" (BURP 2021). Specifically, the cooking center has a production capacity of 1590 meals per production cycle, viz. 1590 meals at lunch and 1590 meals at dinner, and in the research its production share was: (i) 990 meals produced according to the cook-and-chill method per production cycle per day; and (ii) 600 meals produced according to the

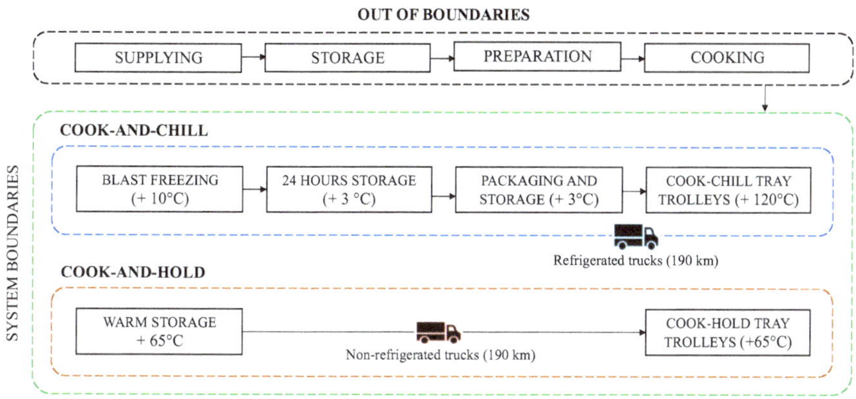

Fig. 3.2 Cook-and-chill versus cook-and-hold foodservice (system boundaries). Retrieved from Bux et al. (2023)

cook-and-hold one. Precisely, primary data were retrieved from the latest energy analysis developed according to the ISO 50001:2018 on the "Energy Management Systems—Requirements and guidelines for use" and from the technical specification of the vehicles (and instruments) used to transport and serve meals, namely: (a) insulated vehicle with MASTER tail lift (Renault) for transportation; (b) single and with double fridge thermal trolleys with customized trays (Burlodge) for transporting and service meals in the different hospital wards; and (c); porcelain dishes, steel cutlery, unbreakable glass for final consumption. Further data were gathered by using an integrated computer system, so-called NOVA (https://www.novasrl.net/) that relies on a cloud-based technology. NOVA offers a specific service, defined "E.trace" that includes information on the receipt of goods and storage, the warehouse withdrawals, the production and packaging, as well as the distribution. In particular, E-Trace provided data related to: (i) delivery traceability via electronic shipping documents; (ii) traceability of food trolleys; (iii) personalized tray traceability to the bed of the patients; (iv) polibox traceability; and (v) traceability of the transported meals. The energetic baseline was developed according to primary data related to the timeframe July 2021–June 2022, in line with the period of administration of the customer satisfaction analysis (see Sect. 3.3.2.2).

The energy consumption analysis is based on the subsequent variables: (i) kilometers traveled by trucks, namely 190 km from the cooking center to the different hospital units; (ii) hours of functioning of the different instruments, for example hours of functioning of the blast freezing or of the cold rooms; (iii) load factors of trucks, storage rooms, and trolleys; and (iv) meals capacity per truck, per blast freezing, per storage room and the trolley. Table 3.1 identified the main variables object of the research considering the baseline scenario, viz. 990 meals prepared with the cook-and-chill method and 600 meals prepared with the cook-and-hold one.

Table 3.1 Baseline scenario for the energy consumption analysis

No.	Cook-and-chill method	Hours or km	Unit cons.[a]	Load factor (%)
1.	Blast freezing (+ 10 °C)	7.50 h	6.40 kWh	70
2.	Storage in cold rooms (+ 3 °C)	24.00 h	7.50 kWh	30
3.	Energy cons. to keep temperature (+ 3 °C)	4.00 h	7.50 kWh	30
4.	Trans. with refrigerated trucks	190 km	7 km/L	100
5.	Use of cook-and-chill trolleys (+ 120 °C)	0.83 h	8.00 kWh	85
No.	**Cook-and-hold method**	**Hours or km**	**Unit cons.[a]**	**Load factor (%)**
1.	Trans. with non-refrigerated trucks (+ 65 °C)	190 km	12 km/L	100
2.	Use of cook-and-hold tray trolleys (+ 65 °C)	0.25 h	6.20 kWh	40

Note Trans. = transportation; cons. = consumption; cap. = capacity; req. = requirement
[a]The "Unit cons." refers to the liters or electricity consumption of each vehicle or each facility. For instance, considering each vehicle contains 180 meals, six vehicles are required to transport meals (i.e., 990/180 = 5.50 = 6 vehicles). The values in the Table consider the km traveled by all vehicles (identified by their license plate) per month, the average quantity of diesel consumed by each vehicle, and the number of meals transported

In the research, the energy consumption analysis refers either to electricity consumption (i.e., kWh) or diesel consumption (i.e., liters) and, as to compare these results, also to tons of oil equivalent (toe).

3.3.2.2 Customer Satisfaction Analysis

The questionnaire used in the case-study was developed in line with the documents provided by the Regione Puglia and was approved by the Department of Health Promotion, Social Wellbeing and Sport, Section Promotion of Health and Wellbeing, Service Food Safety and Veterinary Health (Ministry of Health 2022). It was administered to all patients hospitalized in the ASL Bari on a weekly basis, for an amount of 984 respondents, from November 22 to November 29, 2021, keeping their identity anonymous. In terms of questions, the questionnaire was composed of both qualitative and quantitative questions, so as to comprehend as much as possible of the foodservice in terms of quality of meals and consumer behavior. Respondents were asked to evaluate their qualitative perception of meals at lunch and at dinner, with specific regard to the main course, the second course and the side dish, using a 3-Likert scale (i.e., 1 = bad quality; 2 = average quality; 3 = good quality). In terms of food waste drivers, defined in the questionnaire as "reasons for disliking", patients had the chance to identify: bad taste, bland, overcooked, little seasoned, "al dente", "cold" or salty, at already presented in previous studies on the topic (Piciocchi et al. 2022; Donini et al. 2008).

In practice, questionnaire were administered to patients by researchers and experts in the field (i.e., hospital dietary staff) through the use of tables, and results were catalogued in the NOVA system that is a user-friendly management software and presents a transparent and clear dashboard. The adoption of the E.trace software from NOVA offered the chance to track meals and to distinguish among those served according to a cook-and-chill method and those served according to a cook-and-hold one.

3.3.2.3 Food Waste Analysis

The measurement of food waste was carried out according to the questionnaire method, as discussed by the Commission Delegated Decision 1597/2019 (Official Journal of the European Union 2019). So as to obtain clear figures and allow respondents to use an intuitive method for measuring, each patient was asked to identify how much food waste consumed (and so, how much food was left in the plate) according to the subsequent scale: (i) the entire dish (i.e., no leftovers); (ii) three-quarter of a dish (i.e., 25% of food waste); (iii) half a dish (i.e., 50% food waste); and (iv) no consumption (100% food waste). These quantities were graphically identified in the questionnaire through a circle composed of for equal segments. In terms of weight, it was considered that each course (i.e., first course, second course, side dish) weights on average 0.20 kg for an amount of about 0.60 kg per meal (i.e., lunch, dinner).

3.4 Results and Discussion

3.4.1 Energy Consumption Results

The energy consumption for the production of 990 meals with the cook-and-chill method is about 0.19 toe, distinguishing between 282 kWh of electricity and 162 L of diesel, whereas the energy consumption for the production of 600 meals with the cook-and-hold method is about 0.08 toe, as follows: 12 kWh of electricity and 63 L of diesel. In terms of energy consumption per meal, which is a rather comparable and representative result, is results that one meal produced with the cook-and-chill method consumes about 0.29 kWh and 0.16 L of diesel, whereas according to the cook-and-hold method it required about 0.05 kWh and 0.15 L of diesel. Table 3.2 illustrates the main results for the energy consumption analysis in the baseline scenario, comparing the current scenario that considers a production of 990 meals with the cook-and-chill method and 600 meals with the cook-and-hold one.

Table 3.3 illustrates the results in the case of which in the cooking center, whose capacity is 1590 meals per production cycle, meals are entirely produced according to the cook-and-chill and the cook-and-hold method.

On a comparable basis, the current case-study highlights that the cook-and-hold method is more efficient in terms of energy compared to the cook-and-chill method, since: (i) diesel consumption could be reduced by 42%, from 244 to 142 L; and (ii) electricity consumption could be reduced by 93%, from 492 to 32 kWh due to the

Table 3.2 Energy efficiency analysis in the baseline scenario

No.	Cook-and-chill method ($n = 990$ meals)	Daily cons.[a]	Daily cons. (toe)
1.	Blast freezing (+ 10 °C)	33.60 kWh	0.006
2.	Storage in cold rooms (+ 3 °C)	54.00 kWh	0.01
3.	Energy cons. to keep temperature (+ 3 °C)	9.00 kWh	0.002
4.	Trans. with refrigerated trucks	162.86 L	0.14
5.	Use of cook-and-chill trolleys (+ 120 °C)	186.25 kWh	0.036
	Total (1 + 2 + 3 + 4 + 5)		0.19
No.	**Cook-and-hold method**	**Daily cons.[a]**	**Daily cons. (toe)**
1.	Trans. with non-refrigerated trucks (+ 65 °C)	63.33 L	0.08
2.	Use of cook-and-hold tray trolleys (+ 65 °C)	12.40 kWh	0.004
	Total (1 + 2)		0.08

Note Trans. = transportation; cons. = consumption

[a]In the light of the variables included in Table 3.1, the results of the daily consumption are given for the diesel consumption by applying the subsequent equation: "(190 km/12 km per L) × 100% × 4 units", or for the blast freezing electricity consumption, as follows: "(7.50 h × 6.40 kWh × 70% × 1 unit)". The meal capacity (i.e., unit) is expressed in terms of meals; for instance, one vehicle transports about 180 meals, whereas one blast freezing facility contains 1000 meals. The units required refer to the number of vehicles or facilities required to supply meals, namely 990 cook-chill meals per production cycle and 600 cook-hold meals per production cycle

Table 3.3 Energy efficiency analysis in the theoretical scenario

No.	Cook-and-chill method ($n = 1590$ meals)	Daily cons.	Daily cons. (toe)	Percentage (%)
1.	Blast freezing (+ 10 °C)	67.20 kWh	0.01	4
2.	Storage in cold rooms (+ 3 °C)	108.00 kWh	0.02	7
3.	Energy cons. to keep temperature (+ 3 °C)	18.00 kWh	0.00	1
4.	Trans. with refrigerated trucks	244.29 L	0.20	69
5.	Use of cook-and-chill trolleys (+ 120 °C)	299.13 kWh	0.06	19
	Total (1 + 2 + 3 + 4 + 5)		0.30	100
No.	**Cook-and-hold method ($n = 1590$ meals)**	**Daily cons.**	**Daily cons. (toe)**	**Percentage (%)**
1	Trans. with non-refrigerated trucks (+ 65 °C)	142.50 L	0.12	95
2	Use of cook-and-hold tray trolleys (+ 65 °C)	32.86 kWh	0.01	5
	Total (1 + 2)		0.12	100

Note Trans. = transportation; cons. = consumption

absence of several energy consuming activities, such as blast freezing, storage in cold rooms and use of cook-and-chill trolleys. Of course, by comparing Table 3.2 with Table 3.3, it is also possible to comprehend the importance of the economies of scale, since passing from a production of 600 meals to a production of 1590 meals in the cook-and-hold method allows for additional energy savings. Specifically, producing meals at the highest capacity of the cooking center allows for a consumption of 0.09 L of diesel and 0.02 kWh per meal in the cook-and-hold method, whereas at less than half capacity the energy consumption per meal increased up to 0.15 L of diesel and 0.05 kWh of electricity. In terms of toe, it results that the overall energy reduction by substituting the cook-and-chill service with cook-and-hold-serve method is about 56%.

Best practices for further reducing energy consumption in the cooking center could either regard electricity or natural gas. In terms of electricity, a relamping intervention could reduce the energy consumption by substituting fluorescent with LED tubes while maintaining the existing ceiling lights, with an additional expected reduction of about 40% and a payback period of no more than four years. Moreover, in terms of cold rooms, education and awareness strategies among the kitchen staff could reduce the energy consumption associated with doors left open during loading and unloading activities of foodstuffs. In terms of natural gas reduction, possible strategies could regard the installation of photovoltaic "solar thermal" systems for the production of domestic hot water, as well as the use of a condensing boiler, which could help reduce energy consumption by 25%. Last, as regards transportation vehicles, some obsolete trucks could be substituted with more environmentally friendly ones.

3.4.2 Customer Satisfaction Results

The results from the questionnaire, which are rather cost-efficient and reliable according to the sample size (i.e., the entire population hospitalized in the nine different hospitals from November 22 to November 29, 2022), reveal the main drivers for food waste. On average, the meal consumption rate is about 92.72% for lunch and 88.70% for dinner, with dry breakfast being the most consumed meal (100%), followed by dessert (93.5%) and second course (89.8%). In terms of food consumption (and food waste) drivers, respondents highlighted that 79.41% of meals were of good quality, while 16.78% and 3.81% or average and bad quality, respectively. Specifically, the main cause for food waste is that "food has a bad smell" (about 22% of respondents), but also "food is bland" (about 19%) and food is overcooked (about 17%). By comparing cook-and-chill and cook-and-hold methods, the highest rate of consumption is associated with the cook-and-hold (97%), whereas it is rather lower for the cook-and-chill method (88%).

However, the interpretation of food consumption behavior among hospital patients is somewhat complex, considering that several variables should be taken (and were not) into account, such as the hospitalization time, the state of health, the ongoing therapies, the personal taste, etc. (Naithani et al. 2008).

3.4.3 Food Waste Results

Table 3.4 summarizes the main results related to the food waste analysis, first distinguishing per day of the ween, then per meal (i.e., lunch and dinner) and subsequently by cook-and-chill and cook-and-serve method.

Considering the results of the customer satisfaction analysis, cook-and-hold method helps reduce food up to 2.54%, whereas in terms of cook-and-chill method is upsurges up to 11.16%. On average, food waste is about 4.29% at lunch and 1.61% at dinner for the cook-and-hold method, whereas its amount raises up to 10.47% and 13.54% in the cook-and-chill method, respectively. Compared to Italian figures, these numbers are rather lower with those identified by the CAMs (GURI 2020), which estimated an amount of food waste at hospital of about 30%. Further details in the field of food waste in the hospital food waste are presented in Chap. 4.

Table 3.4 Food waste results

Day	Meal	Mean C&H	FW/Meal/Day C&H	Mean C&C	FW/Meal/Day C&C
Monday	L	1.38%	0.008 kg	12.80%	0.077 kg
	D	0.00%	0.000 kg	13.59%	0.082 kg
Tuesday	L	1.79%	0.011 kg	13.26%	0.080 kg
	D	2.01%	0.012 kg	29.03%	0.174 kg
Wednesday	L	–	–	–	–
	D	–	–	–	–
Thursday	L	3.22%	0.019 kg	10.6%	0.064 kg
	D	0%	0.000 kg	6.11%	0.037 kg
Friday	L	6.75%	0.041 kg	12.44%	0.075 kg
	D	5.26%	0.032 kg	16.53%	0.099 kg
Saturday	L	1.26%	0.008 kg	19.45%	0.117 kg
	D	0.00%	0.000 kg	6.95%	0.042 kg
Sunday	L	12.50%	0.075 kg	13.10%	0.079 kg
	D	–	–	10.38%	0.062 kg
Total		–	0.205 kg	–	0.985 kg

Note CH. = cook-hold; CC. = cook-chill; FW = food waste. Food waste data distinguish between daily meals and are aggregated per site per method. "Mean CH." identifies the weighted average of food waste recorded in all sites that use the cook-hold method, whereas "Mean CC." identifies the weighted average of food waste recorded in all sites that use the cook-hold method. "FW/meal/day" illustrates the amount of food waste generated per meal per day. The total amount represents the amount of food waste generated per patient per week

3.5 Conclusions

The chapter tried to provide a comprehensive, although preliminary, assessment of energy efficiency, customer satisfaction and food waste measurement in the hospital foodservice on a sample of 984 patients over a one-week period. Specifically, primary data were retrieved from a cooking center in Southern Italy with a production capacity per cycle (i.e., lunch, dinner) of 1590 meals per day, comparing two alternative food-service methods: on the one side, cook-and-chill; on the other side, cook-and-hold. Considering the energy efficiency as the "ratio or other quantitative relationship between results in terms of performance, services, goods or energy, and energy input" and taking care of energy as "electricity, fuel, steam, heat, compressed air and other similar means" (Naithani et al. 2008) during the different foodservice activities from gate-to-cradle, the research highlighted some possible practices to enhance sustainability. Specifically, results defined the environmentally friendly characteristics of the cook-and-hold method compared to the cook-and-chill method, which is rater less energy intensive and more preferred by consumers. The cook-and-hold method, which do not require any blast freezing (at + 10 °C), storage in cold rooms (at + 3 °C), transportation with refrigerated trucks or use of cook-and-chill-trolleys (at + 120 °C), allows for the reduction in diesel consumption by 42% and in electricity consumption by 93%, making it possible to consume 0.09 L of diesel and 0.02 kWh of electricity per meal. In addition, considering the sample, meals served with the cook-and-hold method are more likely to be appreciated by patients, reducing food waste up to 2.54% compared to meals served with the cook-and-chill method, whose food waste rate is about 11%.

However, the case-study presents some limitations. It does not account some variables, which may affect food waste rates among hospital patients, such as the hospitalization time, the state of health, personal tastes, or the ongoing therapies of the sampled patients.

References

Accredia. ISO Survey 2022: Italia 1° in Europa e 2° al mondo per numero di certificati. Access date 19 June 2024. Available from: https://www.accredia.it/2023/11/27/iso-survey-2022-italia-1-in-europa-e-2-al-mondo-per-numero-di-certificati/

Antón-Peset A, Fernandez-Zamudio M-A, Pina T (2021) Promoting food waste reduction at primary schools. A case study. Sustainability 13:600

Arriz-Jorquiera M, Acuna JA, Rodríguez-Carbó M, Zayas-Castro JL (2024) Hospital food management: a multi-objective approach to reduce waste and costs. Waste Manag 175:12–21

Batlle-Bayer L, Bala A, Aldaco R, Vidal-Monés B, Colomé R, Fullana-i-Palmer P (2021) An explorative assessment of environmental and nutritional benefits of introducing low-carbon meals to Barcelona schools. Sci Total Environ 756:143879

Berniak-Woźny J, Rataj M (2023) Towards green and sustainable healthcare: a literature review and research agenda for green leadership in the healthcare sector. Int J Environ Res Public Health 20:908

Boschini M, Falasconi L, Cicatiello C, Franco S (2020) Why the waste? A large-scale study on the causes of food waste at school canteens. J Clean Prod 246:118994
Bouley T, Roschnik S, Karliner J, Wilburn S, Slotterback S, Guenther R, Orris P, Kasper T, Platzer BL, Torgeson K (2017) Climate smart health care. World Bank Group, Washington, United States
BURP (2021) Deliberazione della Giunta Regionale 22 dicembre 2020, n. 2108. Linee di Indirizzo regionali per la ristorazione Ospedaliera. Access date 19 June 2024. Available from: http://www.aforp.org/riservata/trasparenza/delibere/2020/uploads/6a41b_DELIBERAZIONE%20DELLA%20GIUNTA%20REGIONALE%2022%20dicembre%202020,%20n.%202108.pdf
Bux C, Zizzo G, Amicarelli V (2023) A combined evaluation of energy efficiency, customer satisfaction and food waste in the healthcare sector by comparing cook-hold and cook-chill catering. J Clean Prod 429:139594
Chawla G, Lugosi P, Hawkins R (2022) Factors influencing hospitality employees' pro-environmental behaviours toward food waste. Sustainability 14:9015
Cook N, Collins J, Goodwin D, Porter J (2022) A systematic review of food waste audit methods in hospital foodservices: development of a consensus pathway food waste audit tool. J Hum Nutr Diet 35:68–80
Dias-Ferreira C, Santos T, Oliveira V (2015) Hospital food waste and environmental and economic indicators—a Portuguese case study. Waste Manag 46:146–154
Donini LM, Castellaneta E, De Guglielmi S, De Felice MR, Savina C, Coletti C, Paolini M, Cannella C (2008) Improvement in the quality of the catering service of a rehabilitation hospital. Clin Nutr 27(1):105–114
Environmental and Energy Study Institute. Energy efficiency. Access date 19 June 2024. Available from: https://www.eesi.org/topics/energy-efficiency/description
Eurostat (2022) Circular economy. Monitoring framework. Access date 15 June 2024. Available from: https://ec.europa.eu/eurostat/web/circular-economy/monitoring-framework
GURI (Gazzetta Ufficiale della Repubblica Italiana). Servizio di ristorazione e fornitura di derrate alimentare (approvato con DM n. 65 del 10 marzo 2020, in G.U. n. 90 del 4 aprile 2020). Access date 19 June 2024. Available from: https://www.mase.gov.it/sites/default/files/archivio/allegati/GPP/2021/cam_ristorazione.pdf
ISO (2024a) ISO 50001. Energy management. Access date 19 June 2024. Available from: https://www.iso.org/iso-50001-energy-management.html
ISO (2024b) ISO 59020:2024. Circular economy. Measuring and assessing circularity performance. Access date 19 June 2024. Available from: https://www.iso.org/standard/80650.html
Karliner J, Slotterback S, Boyd R, Ashby B, Steele K (2024) Health care's climate footprint. How the health sector contributes to the global climate crisis and opportunities for action. Access date 19 June 2024. Available from: https://noharm-global.org/sites/default/files/documents-files/5961/HealthCaresClimateFootprint_092319.pdf
Knezevic B, Kurnoga N, Anic I-D (2019) Typology of university students regarding attitudes towards food waste. Br Food J 121(11):2578–2591
Li J, Li W, Wang L, Jin B (2021) Environmental and cost impacts of food waste in university canteen from a life cycle perspective. Energies 14:5907
Lorenz-Walther BAS, Langen N (2020) Sustainable changes in a worksite canteen: an exploratory study on the acceptance of guests. J Clean Prod 259:120737
Malefors C, Sundin N, Tromp M, Eriksson M (2022) Testing interventions to reduce food waste in school catering. Resour Conserv Recycl 177:105997
Marques C, Lima JPM, Fialho S, Pinto E, Baltazar AL (2022) Impact of a food education session on vegetables plate waste in a Portuguese school canteen. Sustainability 14:16674
Ministry of Health (2022) Linee di indirizzo nazionale per la ristorazione ospedaliera, assistenziale e scolastica. Access date 19 June 2024. Available from: https://www.salute.gov.it/imgs/C_17_pubblicazioni_3141_allegato.pdf
Naithani S, Whelan K, Thomas J, Gulliford MC, Morgan M (2008) Hospital inpatients' experiences of access to food: a qualitative interview and observational study. Health Expect 11:294–303

NHS (2020) Delivering a 'Net Zero' National Health Service. Access date 19 June 2024. Available from: https://www.england.nhs.uk/greenernhs/wp-content/uploads/sites/51/2020/10/delivering-a-net-zero-national-health-service.pdf

Official Journal of the European Union. Regulation (EC) No. 853/2004 of the European Parliament and of the Council of 29 April 2004. Access date 19 June 2024. Available from: https://eur-lex.europa.eu/legal-content/EN/TXT/PDF/?uri=CELEX:32004R0853

Official Journal of the European Union. Commission Delegated Decision (EU) 2019/1596 of 3 May 2019. Access date 19 June 2024. Available from: https://eur-lex.europa.eu/legal-content/EN/TXT/PDF/?uri=CELEX:32019D1597

Page MJ, McKenzie JE, Bossuyt PM, Boutron I, Hoffmann TC, Mulrow CD et al (2021) The PRISMA 2020 statement: an updated guideline for reporting systematic reviews. BMJ 372(71):1–9

Paiva L, Hernández C, Santos D, Garayoa R, García L, Urdangarín C, Vitas AI (2022) Hospital plate waste assessment after modifications in specific dishes of flexible and inflexible food ordering systems. Sustainability 14:15616

Pancino B, Cicatiello C, Falasconi L, Boschini M (2021) School canteens and the food waste challenge: which public initiatives can help? Waste Manag Res 39(8):1090–1100

Piciocchi C, Lobefaro S, Luisi F, Miraglia L, Romito N, Luneia R, Foti S, Mocini E, Poggiogalle E, Lenzi A, Donini LM (2022) Innovative cooking techniques in a hospital food service: effects on the quality of hospital meals. Nutrition 93:111487

Piras S, Righi S, Banchelli F, Giordano C, Setti M (2023) Food waste between environmental education, peers, and family influence. Insights from primary school students in Northern Italy. J Cleaner Prod 383:135461

Pires I, Machado J, Rocha A, Liz Martins M (2022) Food waste perception of workplace canteen users—a case study. Sustainability 14:1324

Qian L, Li F, Cao B, Wang L, Jin S (2021a) Determinants of food waste generation in Chinese university canteens: evidence from 9192 university students. Resour Conserv Recycl 167:105410. https://doi.org/10.1016/j.resconrec.2021.105410

Qian L, Rao Q, Liu H, McCarthy B, Liu LX, Wang L (2022) Food waste and associated carbon footprint: evidence from Chinese universities. Ecosyst Health Sustain 8:1

Sahal Alharbi N, Yahia Qattan M, Haji Alhaji J (2020) Towards sustainable food services in hospitals: expanding the concept of 'plate waste' to 'tray waste.' Sustainability 12:6872

Sherman JD, MacNeill A, Thiel C (2019) Reducing pollution from the health care industry. JAMA 322(11):1043–1044

Snyder H (2019) Literature review as a research methodology: an overview and guidelines. J Bus Res 104:333–339

Tregear A, Aničić Z, Arfini F, Biasini B, Bituh M, Bojović R, Brečić R, Brennan M, Colić Barić I, Del Rio D, Donati M, Filipović J, Giopp F, Ilić A, Lanza G, Mattas K, Quarrie S, Rosi A, Sayed M, Scazzina F, Tsakiridou E (2022) Routes to sustainability in public food procurement: an investigation of different models in primary school catering. J Clean Prod 338:130604

UNI (2022) UNI/TS 11820. Misurazione Della Circolarità—Metodi ed Indicatori per la Misurazione dei Processi Circolari Nelle Organizzazioni. Ministero dell'Ambiente e della Sicurezza Energetica, Rome, Italy

van Eck NJ, Waltman L (2021) VOSviewer manual. Version 1.6.17. Universiteit Leiden, Leiden

Wang F, Shreedhar G, Galizzi MM, Mourato S (2022) A take-home message: workplace food waste interventions influence household pro-environmental behaviors. Resour Conserv Recycl Adv 15:200106

Wang L, Zhu L, Pan S, Wang S (2023) Low-carbon emitting university campus achieved via anaerobic digestion of canteen food wastes. J Environ Manag 335:117533

Xu L, Kong Y, Wei M, Wang Y, Zhang M, Tjahjono B (2021) Combatting medical plastic waste through visual elicitation: insights from healthcare professionals. J Clean Prod 329:129650

Zhao C, Panizza C, Fox K, Boushey CJ, Byker Shanks C, Ahmed S, Chen S, Serrano EL, Zee J, Fialkowski MK, Banna J (2019) Plate waste in school lunch: barriers, motivators, and perspectives of SNAP-eligible early adolescents in the US. J Nutr Educ Behav 51(8):967–975
Zhou H, Yang Q, Gul E, Shi M, Li J, Yang M, Yang H, Chen B, Zhao H, Yan Y, Erdoğan G, Bartocci P, Fantozzi F (2021) Decarbonizing university campuses through the production of biogas from food waste: an LCA analysis. Renew Energy 176:565–578

Chapter 4
Sustainable and Environmentally Friendly Meals

Abstract Foodservice should provide nutritional, safe, and environmentally friendly meals, especially in hospitals. In Italy, the achievement of this aim requires the adoption of the minimum environmental criteria for foodservice. However, the application of these guidelines is necessary, but not sufficient, since it may exist a tension between sustainability of meals and customer satisfaction. This chapter tests a significant hypothesis for foodservice, namely that environmentally friendly diets may lead to greater food waste. It employs the carbon footprint method to measure the environmental sustainability of meals in foodservice hospitals and carries out questionnaires to quantify food waste among 3454 patients over a three-year period. Food waste at lunch is approximately 8.6–13.9%, whereas at dinner it increases up to 18%. The highest peak at lunch is 24.6%, and the highest peak at dinner is about 30.7%. Given that the hypothesis at the foundation of the research has been confirmed, the chapter proposes sustainable menus with the aim of pursuing either environmental sustainability or customer satisfaction.

Keywords Carbon footprint · Customer satisfaction analysis · Food waste measurement · Global warming potential · Environmentally friendly diets · Life cycle assessment

4.1 Introduction

The healthcare sector is essential for promoting human health and social well-being and nowadays, more than in previous years, it aspires to a fairer, healthier, and more environmentally friendly foodservice for its patients, since "food is medicine" and for some pathologies, foods are more effective than drugs on the path to healing (Carino et al. 2020; Downer et al. 2020). In this field, foodservice companies have a significant impact from the economic, social, environmental, and nutritional perspective, being responsible for the preparation, the transportation and the administration of meals to patients. Furthermore, they have the chance to promote, through sustainable and

C. Bux, *Strategies for a Sustainable and Environmentally Friendly Hospital Foodservice Management*, Environmental Humanities: Transformation, Governance, Ethics, Law 6,
https://doi.org/10.1007/978-3-031-82050-2_4

environmentally friendly diets, healthy and inclusive food habits among consumers (Eriksson et al. 2023; de Visser-Amundson 2022).

Scholars and public authorities were highly interested in the environmental impacts of foodservice, and literature on food waste minimization and environmental impacts spread in recent years, especially in nurseries (González-García et al. 2021), schools (Chiaverina et al. 2023; Simon et al. 2023) and universities (Li et al., 2021). In public schools, the average carbon footprint per meal was estimated at 1.9 $kgCO_2eq$ in France (Chiaverina et al. 2023) and 1.30 $kgCO_2eq$ in Spain (Simon et al. 2023). In terms of food waste, Li et al. (2021) evaluated the average carbon footprint of food waste at about 2.18 $kgCO_2eq$ per kg of food waste in China, considering a mixed hydrothermal treatment and aerobic composting. From the methodological perspective, the assessment of the carbon footprint is based on an input–output approach, which is essential for tracking the environmental flows and for identifying the largest causes of emissions, resource consumption and the associated impacts (Fang et al. 2016; Laurent and Owsianiak 2017). However, research on the environmental impacts of hospital foodservice is rather scarce, and none seem to investigate the current hypothesis:

H_1 Environmentally friendly diets, which are composed of meals with a low carbon footprint, may lead to greater food waste, measured in terms of customer satisfaction, among hospital patients.

By testing the hypothesis (H_1), the chapter evaluates the global warming potential of meals and identifies the best combinations of foods from either the environmental, the nutritional or the social perspective. The research reports the results from a three-year comparative assessment in nine hospitals on a sample of 3454 patients. On average, food waste at lunch ranges from 8.6 to 13.9%, with the highest peak at 24.6%, whereas food waste at dinner accounts from 7.0 to 18.4%, with the highest peak at 30.7%. The most wasted foods are eggs, dairy products and pasta. The weekly carbon footprint is 6.9–11.5 $kgCO_2eq$ at lunch and 6.4–7.0 $kgCO_2eq$ at dinner. It provides useful insights for boosting sustainability in the foodservice and defines efficient pathways and intervention for public authorities, food companies and healthcare practitioners toward the reduction of food waste and the increase and nutritive meals.

4.2 Carbon Footprint

Sustainable development is defined as "meeting the needs of the present without compromising the ability of future generations to meet their own needs" and was first introduced in 1987 in the document entitled "Our Common Future", so-called "Brundtland Report" (Brundtland 1987). Nowadays, the concept of sustainability from either economic, social or environmental perspective became a competitive factor for companies to penetrate local and international markets (Tarnovskaya 2023), but also to respond to the urgent requirements from public authorities. Precisely, environmental sustainability asks companies to pay attention to their energy and water consumption and to the carbon emissions associated with products and processes, and in this context some tools were developed to measure carbon emissions, such as the carbon footprint.

The carbon footprint is based on the measurement of greenhouse gases along a product, service or process lifecycle, and assumed great importance for organizations. Its conceptualization dates back to 1996, when it was first developed the concept "ecological footprint" as the amount of productive land or sear required to sustain a given number of humans (Wackernagel and Rees 1996). In recent years, the carbon footprint was defined as the "the quantity of greenhouse gases expressed in terms of CO_2eq, emitted into the atmosphere by an individual, organization, process, product, or event from within a specified boundary" (ISO 2018a, b; Ercin and Hoekstra 2012). However, several greenhouse gases could be released into the atmosphere by anthropic activities, and for this reason it was introduced the global warming potential (GWP) index, defined as their ability of each gas to absorb energy (i.e., radiative efficiency) and the duration to resist in the atmosphere (i.e., lifetime) (United States Environmental Protection Agency 2024). Considering the Kyoto Protocol, some GWP_{100} (i.e., how much energy the emissions of 1 t of a greenhouse gas will absorb over a given period of time) were defined, as follows: (i) carbon dioxide (CO_2) = 1 per kg of emissions; (ii) methane of fossil origin (CH_4) = 29.8; (iii) methane of non-fossil origin = 27.2; (iv) nitrous oxide (N_2O) = 273; (iv) hydrofluorocarbons (HFC_s) = from 4.84 to 14,600; (v) perfluorocarbons (PFCs) = from 0.004 to 12,400; and (vi) sulfur hexafluoride (SF_6) = 25,200 (International Panel for Climate Change 2021, 2006).

In the past, before the standardization of the carbon footprint method at the global scale, different guidelines and protocols were developed to measure greenhouse gas emissions, as follows: (a) the Greenhouse Gas Protocol of WSI/SBCSD (i.e., World Resource Institute at the World Business Council on Sustainable Development) (Greenhouse Gas Protocol 2024); (b) the ISO 14025 on "Environmental labels and declarations—Type III environmental declarations—Principles and procedures", introduced in 2006 and reviewed in 2020 (ISO 2006a); (c) the PAS 2050:2011 on the specification for the assessment of the life cycle greenhouse gas emissions of goods and services and the PAS 2050-1:2021 on the assessment of life cycle greenhouse gas emissions from horticultural products, published by the British Standard Institution (British Standard Institution 2024); (d) the 2006 IPCC (i.e., Intergovernmental Panel on Climate Change) Guidelines for National Greenhouse Has Inventories (Intergovernmental Panel on Climate Change 2006). Furthermore, in addition to international standards, some local guidelines were presented over time, such as those introduced by the Department of Food and Rural Affair (DEFRA) or the Carbon trust in the United Kingdom (Pandey et al. 2021). Nowadays, some of the most used standards for measuring carbon emissions at micro-, meso- and macro-level are the ISO 14067:2018 on "Greenhouse gases—Carbon footprint of products—Requirements and guidelines for quantification" at the product level (ISO 2018b) and the

ISO 14064:2018 for the evaluation of the carbon footprint at the organization level (ISO 2018a).

The measurement of the carbon footprint could be conducted manually or using specific software, such as the Sphera (also known as GaBi), the SimaPro, the Autodesk or Intelex (Ormazabal et al. 2014). Moreover, there are also freeware, such as openLCA, which allows for the assessment of the global warming potential (or climate change, depending on the impact category) for free. However, the overarching point related to carbon footprint assessment regards the data collection, which represents one of the most critical points in the analysis and which was defined as "a research job itself" (OpenLCA Nexus 2024).

Details related to the stepwise approach to carry out a carbon footprint analysis could be found at Sect. 4.3.2.

4.3 Materials and Methods

The research is based on the definition of foodservice presented in the CAMs (GURI 2020), which considers different steps such as preparation of meals, their transportation and administration from the cooking center to the hospital units, and the management of food surpluses (or food waste). Specifically, it applies a stepwise approach, as follows: (i) data collection through a three-years customer satisfaction analysis administered to patients by a questionnaire, which is essential to comprehend food waste quantities; and (ii) carbon footprint assessment from gate-to-crave, including preparation of meals, transportation from the cooking center to the hospital units and food waste disposal. Primary data were collected in nine hospitals located in Southern Italy, and the study area corresponds to that presented in Sect. 3.3.1 (Chap. 3).

4.3.1 Customer Satisfaction and Food Waste Analysis

Based on the Commission Delegated Decision (EU) 2019/1597 on "a common methodology and minimum quality requirements for the uniform measurement of levels of food waste" (Official Journal of the European Union 2019), the questionnaire represents an appropriate method for gathering both qualitative (i.e., food waste drivers) and quantitative (i.e., food waste quantities) information among final consumers, i.e., hospital patients. In the present case-study, as outlined by previous studies on the topic, the questionnaire helps measure the efficiency of the foodservice company and the needs and expectations of the hospital patients (Alaimo et al. 2022).

The questionnaire was developed and approved by the Department of Health Promotion, Social Wellbeing and Sport, Section Promotion of Health and Wellbeing, Service Food Safety and Veterinary Health of the Regione Puglia (Ministry of Health

2022) and was administered to all patients hospitalized in the nine units on a three-years period, as follows: (i) from November 22 to November 29, 2021, for an amount of 984 respondents; (ii) from October 23 to October 30, 2022, for an amount of 1348 respondents; and (iii) from September 24 to October 1, 2023, for an amount of 1122 patients. Anonymous interviews were carried out by researchers and hospital dietary staff through tables, and primary data were catalogued by the NOVA software (see Sect. 3.3.2.2, Chap. 3), which provides users with flexible and modular data in a transparent dashboard.

In terms of qualitative variables, the questionnaire asked respondents to define: (i) their satisfaction rating, to be identified through a sad, neutral or happy smile; and (ii) the reasons for disliking (i.e., bad taste, bland food, overcooked food, little seasoned food, "al dente", cold food, salty food), in line with previous studies on the topic (Piciocchi et al. 2022; Donini et al. 2008). In terms of food waste measurement, respondents were asked to identify the quantity of food consumed, as already defined in Sect. 3.3.2.3 (Chap. 3), by distinguishing per meal (i.e., lunch, dinner) and per specific course (i.e., first course, second course, side dish).

4.3.1.1 Composition of the Menus

In light of the CAMs for the hospital foodservice (GURI, 2020), menus are composed of: (i) at least 40% of organic foods for adults and 50% for pediatric and adolescent users; (ii) at least 60% of organic fruit and vegetables. In addition, as foreseen by the CAMs, menus: (iii) contain low animal protein; (iv) respect seasonality; (v) do not use any pre-processed and fifth range foods; and (vi) are composed of local foods (and zero-km foods), if possible. Precisely, the application of the CAMs contributes to the adoption of environmental technologies and environmentally friendly products and processes among companies, as to leverage the market and boost sustainable performances among economic operators (Ministero dell'Ambiente e della Sicurezza Energetica 2021). In Italy, the effectiveness of the CAMs was ensured through the article 18 of the Law 221/2015 and, subsequently, through the article 34 containing "Energy and environmental sustainability criteria" of the Legislative Decree 50/2016 "Procurement Code" (amended by Legislative Decree 56/2017), which made its application mandatory by all contracting stations (i.e., foodservice companies).

4.3.2 Carbon Footprint Assessment

In light of the hypothesis H_1 (i.e., environmentally friendly diets, which are composed of meals with a low carbon footprint, may lead to greater food waste, measured in terms of customer satisfaction, among hospital patients), one core aspect of the current case-study is to determine the environmental sustainability of meals through their carbon footprint assessment. Precisely, this assessment was conducted by considering the ISO/TS 14067:2018 on "Greenhouse gases—Carbon footprint of products—Requirements and guidelines for quantification" at the product level (ISO 2018b), with the purpose of estimating the greenhouse gas emissions associated with meals, including raw ingredients, meals preparation, foodservice (i.e., logistics), transportation and food disposal. The assessment is based on the $GWP_{100,}$ which is a measure that converts the direct and indirect emissions along the lifecycle into CO_2eq over a period of 100 years (Rana et al. 2023).

The carbon footprint assessment follows the life cycle assessment stepwise approach, as outlined by the ISO 14040:2006 (ISO 2006c) and the ISO 14044:2006 (ISO 2006b): (i) definition of the goals and scope of the assessment, with specific regard to the identification of the functional unit (i.e., "a quantity of a product or product system on the basis of the performance it delivers in its end-use application") (DOE Advanced Manufacturing Office 2022) and the system boundaries; (ii) the life cycle inventory; (iii) the impact assessment, with regards to the global warming potential (expressed in CO_2eq emissions); and (iv) interpretation of results.

The goal of the case-study, for the sake of testing H_1, is to estimate the carbon footprint of three different menus, as outlined in Table 4.1 for lunch and in Table 4.2 for dinner. These tables distinguish among "least consumed" and "most consumed" meals during the three years (i.e., from 2021 to 2023) meals according to the customer satisfaction analysis. In the analysis, "least consumed" meals are those combinations of first course, second course and side dish that recorded the lowest satisfaction rate among consumers, whereas "most consumed" foods are those combination that recorded the highest satisfaction rate. Menus vary on a weekly basis (from Monday to Sunday) and are composed of first course, second course and side-dish.

The functional unit is represented by "one-week menu at lunch" and "one-week menu at dinner", distinguishing among first course, second course and side dish. The system boundaries consider a gate-to-cradle approach and take care of agricultural processed foods, animal husbandry, fishing and crops. Moreover, considering the environmental implications of logistics, the carbon footprint evaluates the carbon emissions associated with transportation from the cooking center to the different hospital units according to the cook-and-chill method as described in Chap. 3, namely: (i) blast freezing; (ii) 24-hourse storage; (iii) transportation with refrigerated trucks for about 190 km; and (iv) on-site service. Last, as concerns the waste management pathway, the baseline scenario measure the carbon emissions associated with landfilling of food waste, but alternative scenario evaluate the emissions associated with incineration and composting, as well as the environmental burdens associated with the cook-and-hold method.

Table 4.1 Menus composition at lunch

Day	Course	"Most consumed" menu (composition)	"Least consumed" menu (composition)
Mon	1st	Pasta (100 g), zucchini (40 g), pork ham (10 g)	Pasta (100 g), tomato (40 g), olive (10 g)
	2nd	Beef hamburger (120 g)	Chicken (120 g)
	Side	Peas (120 g), pork ham (10 g)	Spinach (200 g)
Tue	1st	Pasta (100 g), tuna (40 g)	Pasta (100 g), tomato (40 g)
	2nd	Swordfish (150 g)	Swordfish (150 g)
	Side	Green beans (200 g)	Chard (200 g)
Wed	1st	Rice (100 g), tomato sauce (4 g)	Pasta (40 g), vegetables (60 g)
	2nd	Turkey (120 g)	Turkey (120 g)
	Side	Eggplant (200 g)	Eggplant (200 g)
Thu	1st	Pasta (100 g), bechamel (50 g), tomato sauce (50 g)	Pasta (100 g), cheese (10 g)
	2nd	Cod (150 g)	Chicken (120 g)
	Side	Potatoes (200 g)	Carrots (200 g)
Fri	1st	Pasta (100 g), bechamel (50 g), tomato sauce (50 g)	Pasta (100 g), tomato (40 g)
	2nd	Cod (150 g)	Semi-mature cheese (100 g)
	Side	Vegetables (200 g)	Chard (200 g)
Sat	1st	Pasta (40 g), meat (60 g)	Rice (40 g), vegetables (60 g)
	2nd	Turkey (120 g)	Pork ham (60 g)
	Side	Chard (200 g)	Potatoes (200 g)
Sun	1st	Pasta (100 g), bechamel (50 g), tomato sauce (50 g)	Pasta (100 g), tomato (40 g)
	2nd	Pork (120 g)	Peas (120 g), pork ham (10 g)
	Side	Lettuce (50 g)	Fennels (200 g)

Figure 4.1 illustrates the system boundaries of the carbon footprint assessment.

Primary data were retrieved according to the customer satisfaction analysis (see Sect. 4.3.1) and were analyzed according to the mass-balance principle (i.e., material flow analysis of either food and energy inputs or waste outputs). In terms of emission factors of food production and electricity and diesel consumption, data were retrieved from the SU-EATABLE LIFE database (Petersson et al. 2021) and from research articles (Benvenuti et al. 2022; Volanti et al. 2022; Kolbe 2020). The impact assessment followed the subsequent Eq. (4.1):

$$CF_{TOT.} = \sum_{i=1}^{n}\left[\overline{CF}_i + (FW_i \times EF_{EOL})\right] + \delta \tag{4.1}$$

Table 4.2 Menus composition at dinner

Day	Course	"Most consumed" menu (composition)	"Least consumed" menu (composition)
Mon	1st	Rice (40 g)	Pasta (40 g), vegetables (60 g)
	2nd	Potatoes (150 g), eggs (10 g), semi-mature cheese (10 g)	Semi-mature cheese (100 g)
	Side	Peas (200 g)	Chard (150 g), tomatoes (50 g)
Tue	1st	Rice (100 g), tomatoes (20 g), tuna (20 g), carrots (10 g)	Pasta (40 g), beef meat (60 g)
	2nd	Zucchini (150 g), eggs (10 g)	Pork ham (60 g)
	Side	Lettuce (50 g)	Carrots (50 g)
Wed	1st	Rice (100 g), tomato sauce (40 g)	Peas (50 g), chard (50 g)
	2nd	Plaice (150 g)	Semi-mature cheese (100 g)
	Side	Lettuce (50 g)	Green peas (200 g)
Thu	1st	Lentils (50 g), barley (50 g)	Pasta (100 g), asparagus (20 g)
	2nd	Plaice (150 g)	Semi-mature cheese (100 g)
	Side	Green beans (200 g)	Carrots (50 g)
Fri	1st	Pasta (40 g), vegetables (60 g)	Pasta (100 g)
	2nd	Pork (150 g)	Pork (120 g)
	Side	Fennels (200 g)	Lettuce (50 g)
Sat	1st	Pasta (40 g), lentils (60 g)	Pasta (40 g), potatoes (60 g)
	2nd	Cheese (70 g)	Vegetables (150 g), cheese (20 g), eggs (10 g)
	Side	Lettuce (20 g)	Green beans (200 g)
Sun	1st	Rice (40 g), vegetables (60 g)	Pasta (40 g), vegetables (60 g)
	2nd	Tuna (150 g)	Pork ham (60 g)
	Side	Lettuce (50 g)	Green beans (200 g)

$$\text{with } \overline{\text{CF}}_i = (\text{MF}_i \times \text{EF}_i) \text{ and } \delta = 7x(0.10 + 0.42)\ \text{kgCO}_2\text{eq} \tag{4.1.1}$$

where (i) $\text{CF}_{\text{TOT.}}$ = carbon footprint of the one-week menu at lunch and at dinner; (ii) MF = material flows; (iii) EF = emission factor; (iv) FW = food waste; (v) i = single input considered in the meal production; (VI) EOL = end-of-life pathway; and (vii) δ = coefficient associated with the cook-chill catering and refrigerated transportation from the cooking center to the hospital, calculated on the basis of data presented in Chap. 3.

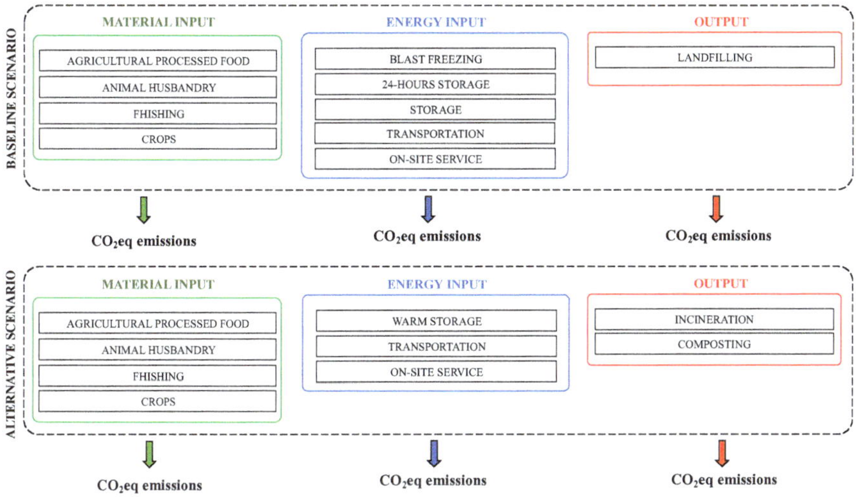

Fig. 4.1 Input, output, and processes included in the carbon footprint analysis

4.4 Results

4.4.1 Results of the Customer Satisfaction Analysis

The current customer satisfaction analysis, carried out among 3454 patients located in nine hospitals, represents one of the first assessments with such a large and representative sample size in food waste hospital research (see Chap. 5). Through qualitative and quantitative questions, the customer satisfaction analysis reports essential information to define sustainable strategies aimed at addressing different challenges, such as the reduction of food waste, the increase of customer satisfaction and the design of environmentally friendly diets. These goals are all included in the hypothesis H_1.

Table 4.3 synthesizes the food consumption rate at lunch and at dinner from Monday to Sunday, distinguishing between first course, second course and side-dish, from 2021 to 2022.

Daily, the most consumed menu was that of Saturday in 2021 (about 95%), followed by the menu of Thursday in 2021 (about 94%) and the menu of Thursday in 2023 (about 90%). On average, the most consumed menus are composed of tastier courses, such as baked pasta or Apulian chops, that are part of the so-called Mediterranean diet (Altamore et al. 2020). The adoption of these courses are more likely to be preferred by patients because entail territoriality and therefore remind respondents to their daily habits, but they also respect environmental sustainability, ethical principles and health protection (Altamore et al. 2017). On the other hand, the least consumed meals encompass pasta in broth, ham and chards. Specifically, in accordance with previous studies on the "emotional eating concept" (Werthmann et al. 2014), broth, ham and semi-mature cheese, as well as vegetables such as chards,

Table 4.3 Food consumption rate at lunch and dinner

Year	Day	Lunch			Dinner			Mean (%)
		FC (%)	SC (%)	SD (%)	FC (%)	SC (%)	SD (%)	
2021	Monday	89.30	90.60	89.70	90.90	88.00	83.00	88.60
	Tuesday	89.30	88.40	87.50	82.40	78.60	74.30	83.40
	Wednesday	–	–	–	–	–	–	–
	Thursday	85.10	92.60	91.70	97.80	97.80	99.40	94.10
	Friday	86.30	88.40	89.20	82.00	89.50	78.40	85.60
	Saturday	91.30	93.90	95.80	96.30	96.60	96.10	95.00
	Sunday	75.40	88.60	84.00	85.50	86.40	91.90	85.30
2022	Monday	84.50	90.30	87.90	85.70	86.90	92.00	88.00
	Tuesday	90.80	92.40	91.40	85.60	89.20	82.20	88.60*
	Wednesday	86.90	90.20	87.50	88.20	81.90	95.10	88.30
	Thursday	88.60	92.00	93.20	83.60	95.50	84.00	89.50**
	Friday	87.90	88.50	87.70	86.10	92.20	96.50	89.80***
	Saturday	97.50	92.40	89.50	85.10	79.30	69.30	85.50***
	Sunday	87.90	96.70	88.40	76.50	92.30	100.00	90.30***
2023	Monday	81.70	89.00	87.10	85.60	91.70	93.10	88.00
	Tuesday	80.30	84.30	84.60	90.60	90.60	91.50	87.00
	Wednesday	80.00	92.00	90.00	79.40	84.60	88.60	85.80
	Thursday	87.20	94.60	90.70	86.00	92.00	93.30	90.50***
	Friday	79.20	89.60	87.80	77.10	90.20	91.20	85.80
	Saturday	90.80	94.20	94.80	70.00	77.40	96.60	87.30***
	Sunday	88.70	92.70	89.60	85.60	92.40	88.80	89.60***

Note FC = first course; SC = second course; SD = side dish. Significance at 0.001 = *, 0.01 = **, 0.05 = ***. Quantities are rounded at the first digit. The *t*-test was used to test differences between the means of 2022 and 2023 with the baseline 2021, to understand whether the differences are statistically significant at $\alpha = 0.05$. The null hypothesis (H_0) states that there is no significant difference between the means of 2022 and 2023 and the mean of 2021, whereas the alternative hypothesis (H_1) assumes that there is a significant difference

could easily generate negative feelings and less enjoyment of meals, with negative effects on food consumption (and positive effects on food waste, in terms of quantity) (Jabeen et al. 2023).

Of course, appealing and tasty food can help patients regain their strength and heal, providing them with normal and daily feelings, but their environmental sustainability (in terms of carbon emissions) must be considered (Murphy 2017).

However, although the food waste percentage per day is of absolute interest, the analysis should go more in-depth, identifying food waste quantities per lunch and dinner and, of course, distinguishing among first course, second course and side dish. On average, lunch is more likely to be appreciated compared to dinner. Over the three years, its consumption rate was about 89.00%, which means that food waste was estimated at 11%. In terms of dinner, its consumption rate as a mean of the three years was about 87.60%, which means that food waste was approximately 12.40%. These results may be attributed to the taste of meals at lunch, which appears higher and more encouraging for patients, whereas at dinner it is not that seasoned. Specifically, at lunch, patients could enjoy meals such as pasta with tomato soup or meat sauce, gratinated fish or meat, while at dinner they are more likely to receive pasta with vegetables, carrots or mixed salads. Considering carbohydrates, final users tend to prefer rice instead of pasta, and the main reason could be linked to the fact that rice holds cooking more than pasta and tends not to overcook easily. In addition, patients prefer some vegetables more than others, such as potatoes and green beans compared to chards, fennels and carrots.

Based on the current data, the most preferred lunch is consumed at 93.40% (which means that food waste is equal to 6.60%) and is composed of pasta with capers and olives, meatloaf (or roast) and green beans. In terms of dinner, the most preferred dinner includes meatloaf and spinach (food consumption = 91.60%, food waste = 8.40%). On the other hand, when looking at the least consumed meals, it results that lunch includes rice with vegetables, ham and fennels (82.70%), whereas the least consumed dinner is based on rice in broth and vegetable pie (that includes vegetables, eggs and cheese).

Getting more in-depth, the most appreciated course is the Apulian chop, which was consumed almost totally (96.70%), followed by the barley soup and the cheese omelets, whereas the least preferred courses were the pasta with vegetables and the pasta with tomatoes (food consumption from 70 to 75%).

Theoretically, the customer satisfaction analysis through the questionnaire-method helps identify both the quantities of food waste and its drivers, namely the reasons for displeasure. Actually, the main drivers are "little seasoned", followed by "bland" and "overcooked". Specifically, with reference to the least consumed course (side dish), it results that about 57% of respondents disliked it because little seasoned, whereas about 30% because it was bland. It is difficult to appreciate the salad without taste or flavor. In the field of the least consumed first course (e.g., pasta with tomatoes, pasta with vegetables), all respondents agreed that it was "overcooked". However, as already mentioned in previous chapters, the reasons why hospital patients tend to generate food waste are different and go beyond the simple "taste". For example, a series of factors and variables intervene, such as hospitalization time, the different pathologies, the medical treatments to which they are subjected, etc. which contribute to increasing (or decreasing) the level of satisfaction.

4.4.1.1 Results of the Food Waste Analysis

Results of the food waste analysis are presented according to the input–output analysis (i.e., material flow analysis) as outlined in Chap. 1 (Sect. 1.4.5.1). To develop easy-to-understand figures, the current chapter identifies food waste flows through the Sankey diagram, which is defined as a flow diagram for a food consumption process in which the quantities of food waste in the various items of the balance are represented by the width of a band (Wilfrid and Peters 1980).

Figure 4.2 illustrates the Sankey diagram of the most and least consumed meals at lunch, distinguishing by food type and according to three main steps, namely: food preparation (i.e., raw ingredients entering the system), food eaten (i.e., food consumption rate) and food waste. Being the functional unit "one-week menu at lunch" (Sect. 4.3.2), the Sankey diagram reports results according to it, so considering food preparation, food consumption and food waste per patient per week at lunch.

As illustrated by Fig. 4.2, a patient provided with the least consumed lunch tend to waste about 50% more compared that a patient provided with the most consumed one. Specifically, over one week, a patient provided with the most consumed meal is likely to waste about 0.29 kg of food, whereas a patient provided with the least consumed one could waste up to 0.43 kg. Daily, hospital patients waste from 0.04 to 0.06 kg of food waste day in the best and in the worst scenario, respectively. The most consumed foods are pasta (from 0.05 to 0.10 kg per patient per week) and vegetables (from 0.12 to 0.24 kg per patient per week).

In terms of dinner, on a weekly basis, patients provided with the most consumed dinner tend to waste about 0.20 kg of food, whereas patients provided with the least consumed one could waste up to 0.43 kg of food. Specifically, the most wasted food categories are eggs, pasta and vegetables.

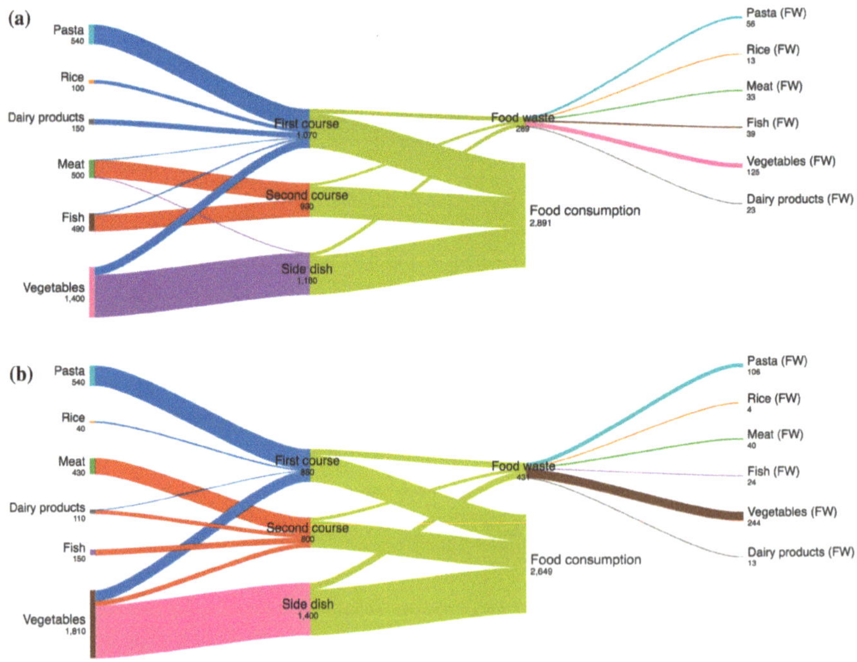

Fig. 4.2 Sankey diagram for food consumption and food waste at lunch, considering the most consumed (**a**) and the least consumed (**b**) food commodities and courses

Under a holistic perspective, food waste at lunch could reach about 14.00%, with its lowest peak at 9.10%, whereas at dinner it could spread up to 18.20%, with its lowest peak at 7.80%.

On the global scale, these results are partially in line with previous literature, which identified higher amounts of food waste. Considering official reports for food waste figures in hospital facilities, Murphy (2017) quantified its amount from 30 to 40% in Canada, whereas it increased up to 40% in the United Kingdom (Barton et al. 2000). These quantities are even higher in the United States, where it was quantified that about 50% of food served to hospital patients is thrown away, but these values tend to be overestimated due to rather small sample sizes, which is not the case for the current research ($n = 3454$). Considering comparable realities in terms of traditions, food qualities and menus, in Italy the CAMs (GURI Gazzetta Ufficiale della Repubblica Italiana 2020) eported an average amount of food waste that is quite in line with the current study (or somehow higher), namely 30%.

Getting more in-depth, previous studies on food waste reported similar quantities, but they suffer from small samples (Chap. 5 will discuss them in detail). For instance, Paiva et al. (2022) reported a daily food waste of about 0.16 kg, as a mean of lunch and dinner, whereas Sahal Alharbi et al. (2020) an amount of 0.15 kg. In the current chapter, the main between lunch and dinner is approximately 12.20%, considering either the most or the least consumed meals.

4.4.1.2 Carbon Footprint Assessment

The carbon footprint assessment reveals interesting results. Considering the functional unit, which is "one-week menu at lunch" and "one-week menu at dinner" per single patient, it results that the carbon footprint is: (i) 11.50 $kgCO_2eq$ for the most consumed lunch; (ii) 6.90 $kgCO_2eq$ for the least consumed lunch; (iii) 6.60 $kgCO_2eq$ for the most consumed dinner; and (iv) 7.00 $kgCO_2eq$ for the least consumed lunch. On the basis of single meals, the highest carbon footprint is recorded at lunch (3.57 $kgCO_2eq$) because it includes meat-based products such as the beef hamburger, whereas the least impacting meal (0.41 $kgCO_2eq$) in mainly plant-based and is composed of fennels and low quantities of pork meat.

Figure 4.3 reports the results of the carbon footprint for the most and the least consumed lunch. Considering the system boundaries of the research, it identified the main hotspots for carbon emissions by distinguishing among processed foods, crops, animal husbandry, fishing, foodservice activities (logistics, i.e., cook-and-chill method), transport for an average of 190 km from the cooking center to the hospital units with refrigerated trucks and landfilling (for the baseline scenario).

On average, a single lunch contributes to climate change with an emission of about 1.65 $kgCO_2eq$ for the most consumed meals and for about 1.00 $kgCO_2eq$ for the least consumed ones. These results match with previous studies on the topic, that highlight how protein-based diets (specifically, meat proteins) have the highest carbon footprint compared to protein-based diets composed of alternative protein sources (Afrouzi et al. 2023; Vidal et al. 2015). In detail, based on Afrouzi et al. (2023), switching from

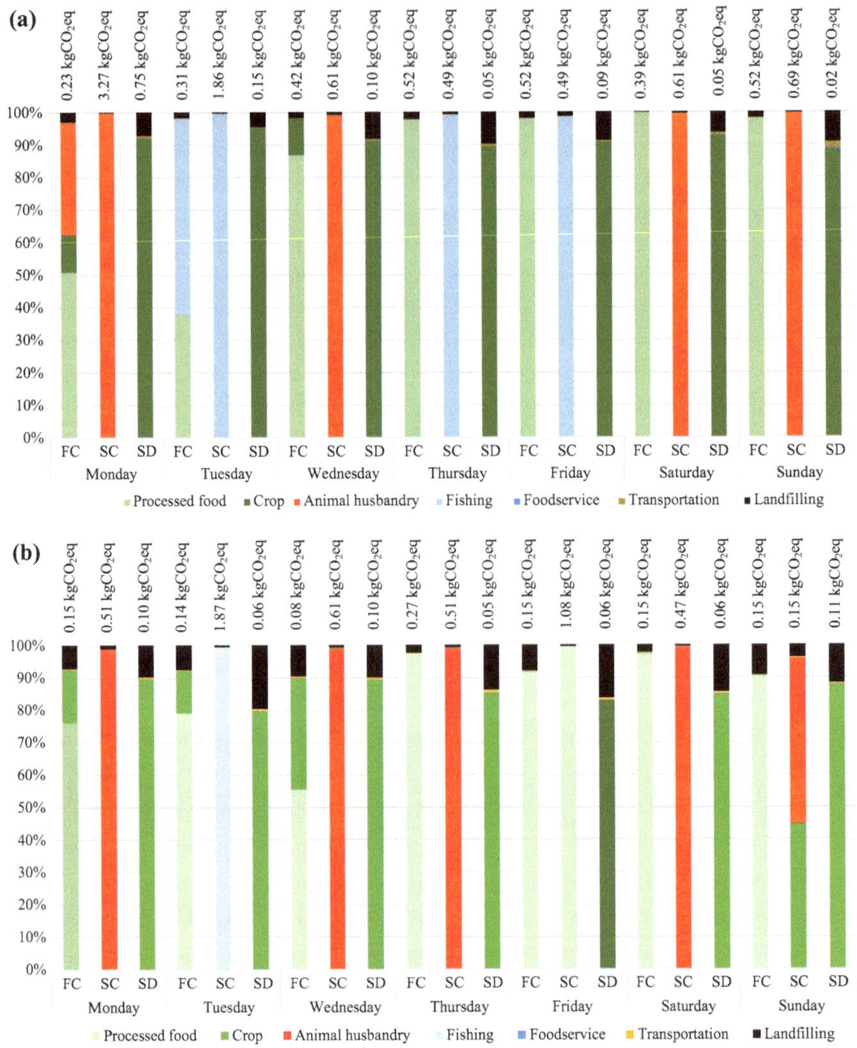

Fig. 4.3 Carbon footprint for the most consumed (**a**) and the least consumed (**b**) lunch

a high-protein diet to a low-protein one could reduce emissions by 48% (considering a diet that include 40 g of proteins) or by 34% (considering a diet that include 60 g of proteins), in the field of meat proteins. Nevertheless, significant reductions could be achieved when substituting meat-proteins with alternative ones (i.e., plant-based proteins) up to 90% (Saget et al. 2021; Khan et al. 2019). In the current case, the top contributors to the carbon footprint at lunch are animal husbandry (about 46%), fishing (26%) and food processing (22%), whereas food waste landfilling account for less than 1%. These values underline the importance of carrying out carbon footprint assessments from cradle-to-grave, in order to identify the main hotspots along the entire agri-food chain, including foodservice activities and logistics (that, in the specific case, account for less than 0.04%).

One average dinner generates from 0.70 to 0.95 kgCO_2eq, with single courses impacting from 0.01 to 0.20 kgCO_2eq. However, considering some outliers, it results that dairy products generate up to 1.10 kgCO_2eq, highly compensated by the low carbon emissions of vegetables. With regards to the categories identified in the current research, animal husbandry reports the highest contribution to climate change (about 32%), then food processing (30%) and fishing (27%). Considering the end-of-life pathway, it results that landfilling impacts up to 2.50%, and this is (also) due to the fact that the amount of food waste generated during dinner is greater than that generated during lunch.

In light of the alternative scenario (Fig. 4.1), the current case-study compares the cook-and-chill with the cook-and-hold method, as well as the landfilling option with composting and anaerobic digestion. These alternatives are likely to generate environmental savings, therefore it appears essential to discuss them more in-depth. Specifically, the environmental impacts of the cook-and-chill method were estimated at 0.10 kgCO_2eq per meal, whereas transportation accounts for about 0.40 kgCO_2eq per meal when considering a full load transportation (i.e., 1800 meals per truck). Switching from the cook-and-chill to the cook-and-hold method can practically reduce emissions associated with electricity consumption from 0.10 kgCO_2eq to about 0.006 kgCO_2eq per single meal. When coming to the diesel consumption, changing from cook-and-chill to cook-and-hold can reduce emissions by 68%, from 0.40 kgCO_2eq to about 0.24 kgCO_2eq per single meal.

In terms of end-of-life pathways, incineration and composting are considered as alternative to landfilling in Italy. Data are retrieved from previous studies on the topic, namely Mondello et al. (2017) for incineration and De Boni et al. (2022) for composting. In line with De Boni et al. (2022), it results that it is possible to reduce carbon emissions by approximately 230% by switching from landfilling to composting. By definition, composting is intended as "the controlled biological decomposition of organic material in the presence of air to form a humus-like material. Controlled methods of composting include mechanical mixing and aerating, ventilating the material by dropping them through a vertical series of aerated chambers, or placing the compost in piles out in the open air and mixing it or turning it periodically" (European Environment Agency 2024). Of course, the percentage of reduction depends on the distance traveled between hospital units and composting area, and in this case-study it is assumed a distance of 15 km/t of food waste. As regards the incineration option, switching from landfilling to incineration is likely to reduce emission due to organic matter decomposition by 51%, but these benefits are highly compensated by the production of energy through energy recovery (Arena et al. 2003). In synthesis "incineration is a method of waste disposal that involves the combustion of waste. It may refer to incineration on land or at sea. Incineration with energy recovery refers to incineration processes where the energy created in the combustion process is harnessed for re-use, for example for power generation. Incineration without energy recovery means the heat generated by combustion is dissipated in the environment" (Eurostat 2024).

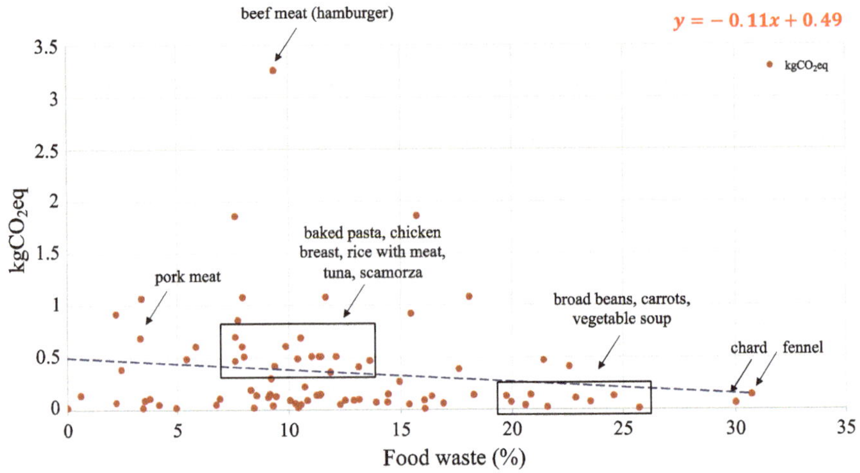

Fig. 4.4 Scatter plot for testing H_1. *Note* The slope coefficient ($-0.11x$) for the best fit line reports a p-value $= 0.01$ at 95%

Soon after the analysis of the alternative scenarios, it is the moment to understand if the hypothesis H_1 (i.e., Environmentally friendly diets, which are composed of meals with a low carbon footprint, may lead to greater food waste, measured in terms of customer satisfaction, among hospital patients) at the basis of the research if confirmed or not. This hypothesis was investigated among the 3454 observations (number of sampled patients) by plotting the carbon footprint of meals (expressed in $kgCO_2eq$) against the quantified food waste rates (expressed in %), as illustrated in Fig. 4.4.

Precisely, as the rate of food waste increases, the amount of carbon emissions decreases, and this confirms H_1. It means that environmentally friendly diets lead to a greater food waste, since the least preferred foods (e.g., beans, vegetable soup, carrots, chards or fennels) have the lowest environmental impact. On the contrary, the food that present the highest rate of consumption (so, the lowest rate of food waste) are those that contribute to climate change the most, such as beef or pork meat. However, although such a rule seem to be confirmed, there are some foods that are either environmentally friendly or appreciated by hospital patients, such as cod or plaice, and this evidence is essential in shaping sustainable diets to achieve both food waste reduction and customer satisfaction.

4.5 Discussion

Shaping sustainable diets that meet both customer satisfaction and nutritional values is rather complex, but this section reports some interesting take-home messages for both practitioners in the sector and researchers in the field of foodservice.

First, when considering alternative foodservice methods, namely cook-and-chill and cook-and-hold, it is essential to comprehend that some foods are suited for a method but some other are not. Precisely, operations such as blast chilling (at + 10 °C) or storage in cold rooms (at + 3 °C) could easily compromise the taste of food and, therefore, the customer satisfaction. In addition to this, there are some foods that hospital patients do not like for clear reasons, such as pasta and tomato that tends to turn out overcooked, or semi-mature cheese or ham that appear unpleasing for aesthetic reasons (Schanes et al. 2018). In this case, the "emotional eating concept" should be considered (Werthmann et al. 2014).

Pasta with tomato sauce is an interesting case, since its carbon emissions are rather low compared to other first courses, including rice with tomato sauce. On average, the emissions associate with pasta with tomato are lower compared to rice with tomato. However, when looking at the environmental sustainability, also food waste quantities must be considered, and it result that pasta generated more waste than rice. For this reason, the amount of food waste offsets the higher impact of the alternatives, so dishes with higher environmental impact are preferable as they generate a lower amount of food waste and fewer emissions. In particular, rice with tomato generates 60% less impact than pasta with tomato in terms of waste disposal and significantly increases customer satisfaction (+ 9%).

In the area of food waste, hospital foodservice must be interpreted differently from households (Hebrok and Boks 2017). In the current study, food waste seems to be a responsibility that is almost exclusively attributed to the organization that takes care of the foodservice. Considering the high impact recorded in the upstream and central stages (e.g., food production, food distribution) compared to the impact in the downstream stages (e.g., landfilling) and considering that the possibilities to change the consumption behavior of hospital patients depend on many variables (e.g., length of hospital stay, health status, personal preferences, ongoing therapies), the role of food service organizations in reducing CO_2eq is essential. One of the most important strategies to reduce food waste could be the analysis of waste trends using questionnaires and surveys (Grant et al. 2023). Regarding cook-chill service, foods that change in flavor when frozen and then brought back to temperature, such as swordfish, must be avoided. Compared to cod, swordfish is less appreciated and has a higher environmental impact (+ 73% per kg of product), which is why it should not be favored over cod (Winther et al. 2009). There are also vegetables that have systematically high rates of food waste, such as chard and carrots.

Possible strategies to reduce greenhouse gas emissions associated with food include replacing high-emission foods with lower-impact foods while maintaining the nutritional properties of meals. In terms of processed foods, pasta is preferable to rice as long as it is possible to avoid overcooking. Dairy products should be avoided as they cause both emissions and waste. Basically, food waste impacts the environment twice: the first time when it is produced and the second time when it is thrown away (Amicarelli et al. 2021). Regarding the environmental impact of crops, which are mainly side dishes, lettuce and green beans are highly valued vegetables compared to carrots, fennel and chard, which are often thrown away. On average, the emissions of crops are similar and vary between 0.25 kgCO_2eq per kg of potatoes and 0.97 $kgCO_2eq$ per kg of asparagus. The variable to pay most attention to in this context is food waste, which contributes significantly to carbon emissions. Finally, among peas and lentils, the former are preferable to the latter, as their choice can reduce emissions by 77%.

The highest emissions come from animal-based foods such as meat and fish. In this context, it is possible to reduce emissions by opting for more environmentally friendly productions, such as chicken or turkey breast (Volanti et al. 2022) compared to beef (Mazzetto et al. 2023). Substituting poultry meat for beef would reduce emissions by 541% and pork by 370%. In addition, boneless pork must be preferred over ham and could reduce emissions by 37% (Petersson et al. 2021). In terms of customer satisfaction, boneless pork and turkey are preferred over pork ham by 8% and 2% respectively. Turkey is tastier than chicken, but is outperformed by beef burgers. However, the degree of preference for hamburgers would not justify an exponential increase in carbon emissions.

Similar considerations can be made for fish. Looking at the three types of fish served (cod, plaice and swordfish), cod is more environmentally friendly than plaice (– 86%) or swordfish (– 276%). Cod is also the most preferred fish in terms of customer satisfaction, achieving an average satisfaction score of 92.11% compared to plaice (91.17%) and swordfish (88.32%). Nevertheless, there is still scope for foodservice companies to support more sustainable aquaculture through sustainable choices (Hammer et al. 2022).

Overall, the weekly carbon footprint is 6.90–11.54 $kgCO_2eq$ at lunch and 6.40–6.99 $kgCO_2eq$ at dinner, which means that an average meal is responsible for 1.32 $kgCO_2eq$ at lunch and 0.95 $kgCO_2eq$ at dinner. Compared to other studies on this topic, the current value is somewhat lower. In Italy, it was estimated at 1.67 $kgCO_2eq$ (Cerutti et al. 2018) and at 1.43 $kgCO_2eq$ (Mistretta et al. 2019) in schools, with the difference due to "healthier" meals served in hospitals, i.e. with balanced carbohydrates, proteins and fibers and low in salt, fats and sugars. Hospital food is specifically designed to meet nutritional needs, health conditions and medical requirements and must be nutritionally balanced (Department of Health 2014).

The environmental impact of foodservice is not only influenced by menu selection, but also by decisions about the type of food, transportation and waste disposal. When comparing cook-chill and cook-hold method (Bux et al. 2023), there is room for improvement through the choice of cook-hold system, although this would raise questions related to some microbiological risks (Bryan 1990) and the flexibility and scale of production. Engelund et al. (2007) have described a limited gastronomic offer and some rigidity in the cook-hold production system. In the best-case scenario, landfilling accounts for about 1% of the total carbon footprint, while in the worst-case scenario it accounts for up to 3%. Last but not least, the reduction of food waste can be achieved by improving service quality and refining the physical environment, which have been assessed as positive variables for increasing customer satisfaction (Le et al. 2023).

Previous studies based on controlled dietary patterns, long-term observations and randomized clinical trials (Benvenga et al. 2022; Willett et al. 2019) also provide some important nutritional considerations to classify our results. Dietary patterns with certain characteristics significantly reduce the risk of serious chronic diseases and improve overall well-being. These characteristics include: mainly plant-based protein sources, fats mainly from unsaturated plant sources and low in saturated fats, carbohydrates mainly from whole grains with limited consumption of refined grains, and sugars less than 5% of total energy. In addition, the dietary recommendations recommend at least five portions of fruit and vegetables (excluding potatoes) per day and a moderate intake of dairy products (Willet et al. 2019). Fish (especially large fish such as swordfish) is an important source of omega-3 polyunsaturated fatty acids, but may also contain contaminants, including heavy metals such as mercury, which is known to be an environmental trigger of autoimmunity and can be measured in the thyroid gland (Benvenga et al. 2022; Willet et al. 2019). In line with current evidence, reducing the consumption of red meat is also important for human health (Beal et al. 2023). In 2015, the International Agency for Research on Cancer stated that red meat is probably carcinogenic to humans, while the consumption of processed meat is classified as carcinogenic to humans and reducing the consumption of these products could play an important role in the prevention of cancer and many chronic diseases (González et al. 2020).

4.5.1 Challenges and Strategies for a Sustainable and Healthy Diet

Sustainable diets are required not only for protecting the environment but also to guarantee public health, since unhealthy diets are responsible for millions of deaths each year. In recent years, public authorities implemented the healthy food procurement and service policies, which are based on some core principles, such as: (i) limitation of the intake of free sugars; (ii) shift from saturated fats to unsaturated fats; (iii) elimination of the industrially produced trans fats; (iv) limitation of sodium consumption and use of iodized salt; (v) increase in the consumption of whole grains, vegetables, fruits, nuts and pulses (World Health Organization 2021). In light of latest statistics, around 8 million deaths are caused every year by unhealthy diets, with particular regard to a high consumption of sodium and salt, sugars and fats that lead to obesity and malnutrition (GBD 2019 Risk Factors Collaborators 2020).

Multiple benefits are associated with a sustainable public food procurement and service policies (e.g., CAMs). The adoption of these policies can have public health, economic, and equity benefits and can increase productivity and educational attainment. In terms of public health, heathy diets are likely to reduce the cardiovascular risk factors, such as hypertension and cholesterol, but can also reduce the risk of cancer, heart disease, stroke, or kidney failure (Micha et al. 2018; Mozaffarian et al. 2018; Driessen et al. 2014). Considering the economic benefits, healthy diets can help public authorities reduce their costs associated with healthcare expenditures, whereas in terms of equity, healthy diets can reduce vulnerability through schools. In addition, these diets could support local agriculture, generating employment opportunities among local communities (Pye-Smith 2014).

Latest statistics reveal that about three-quarters of deaths in the world are represented by the so-called "noncommunicable diseases", among which cardiovascular diseases (i.e., heart disease, stroke), cancer, diabetes and chronic respiratory diseases are accounted (World Health Organization 2019). These diseases could be prevented or delayed by addressing some risk factors, such as behavioral (e.g., unhealthy diets), biological (e.g., obesity, cholesterol) or environmental (e.g., air pollution) factors (World Health Organization 2022a, b). It is important to comprehend that unhealthy diets can take different forms, such as undernutrition, shortfall in essential micronutrients or overnutrition (e.g., obesity) and that, each year, about 8 million people die because of unhealthy food consumption patterns (Global Burden of Disease Collaborative Network 2020; World Health Organization 2022a, b). Considering a recent study on the topic (World Health Organization 2024), an investment of 1 USD on healthy diets can have a return on investment of about 12 USD, highlighting the concept that "saving lives also saves money" (World Health Organization 2022a, b).

There are some protocols that can allow the reduction of diseases through nutrition and which are based on a key mechanism: focused initiatives, enablers, amplifiers. Specifically, focused initiatives are interventions whose aims is to tackle unhealthy consumption patterns or particular risk groups, such as obese people. For instance, these interventions could regard the provision of healthy meals at school, some targeted active travel planning assistance (also called "travelsmart") or the development of healthy working environments for employees, which may include gyms or the promotion of flexible working practices. In terms of enablers, which are ineffective alone but represent an essential item to reinforce interventions, it is important to improve knowledge and education on healthy food consumption patterns, for instance through educational campaign during meals. Last, amplifiers are "key to shirting the system and population profile as a whole". Some examples could be the control of advertising high fat, salt and sugar foods, or the incentive for organizations to report on health outcomes of some foods (Government Office for Science 2024).

4.6 Conclusions

Given the urgent need to tackle the negative environmental impact of foodservice, the results of the current study help to identify some possible strategies for greener supply chain management in hospital foodservice. First, a customer satisfaction analysis was conducted among 3408 patients in nine different hospitals in southern Italy to determine their satisfaction and the amount of food waste generated. Secondly, the carbon footprint of different weekly menus was assessed, following an approach from meal preparation to transportation to the hospital and food waste disposal. Over a period of three years, from November 2021 to October 2023, the data of one week's meals was collected.

The carbon footprint reaches its highest value at lunch (3.57 $kgCO_2eq$) as it contains both processed and animal foods (i.e. beef and pork), while a more sustainable meal at dinner generates around 0.40 $kgCO_2eq$ and contains large amounts of plants (e.g., fennel) and small amounts of pork.

One of the key strategies to combat climate change and minimize food waste is to conduct regular, systematic questionnaires and surveys among customers to assess the evolution of their carbon footprint. The most commonly discarded foods are vegetables such as chard, fennel and carrots, and although their environmental impact is rather low compared to other foods, food waste going to landfill still represents a significant burden on the sustainability of hospital foodservice. In addition, pasta is preferable to rice as long as it is possible to avoid overcooking. Dairy products (e.g., semi-matured cheese) should be avoided as they are less satisfying and generate a large amount of food waste. Finally, animal foods should be reduced as much as possible and some particularly polluting products should be replaced, e.g., beef with pork or turkey and swordfish with cod or plaice.

The current study is limited by the degree of uncertainty regarding the emissions associated with each food. Although the study relies on primary data collected using the questionnaire method and on secondary data contained in official and authoritative databases such as SU-EATABLE LIFE. The research suffers from a degree of uncertainty that does not allow the study to be generalized at an international level, but only in Italy. The questionnaire represents an indirect measurement method for food waste, but it is claimed that it underestimates the amount of food. Nevertheless, the present study provides a solid basis for assessing the environmental impact of dietary changes, particularly in hospital foodservice.

Future research directions include exploring scenarios in which unpopular and/ or carbon-intensive meals are replaced with more sustainable or less wasteful ones to simulate the impact of menu recalibration on carbon footprint. Furthermore, the authors intend to apply material flow costing (ISO 14051:2011) as a management tool to assess the potential environmental and financial benefits of material and energy shifts and to increase economic performance by maintaining high levels of customer satisfaction and reduced environmental impact. Finally, the authors intend to develop a clear and reproducible dashboard for green and sustainable foodservice, as informing and engaging consumers in sustainable food is an essential strategy to mitigate climate change and create fair and environmentally friendly food systems.

References

Afrouzi HN, Ahmed J, Siddique BM, Khairuddin N, Hassan A (2023) A comprehensive review on carbon footprint of regular diet and ways to improving lowered emissions. Results Eng 18:101054

Alaimo LS, Fiore M, Galati A (2022) Measuring consumers' level of satisfaction for online food shopping during COVID-19 in Italy using POSETs. Socio-Econ Plann Sci 82:101064

Altamore L, Bacarella S, Columba P, Chironi S, Ingrassia M (2017) The Italian consumers' preferences for pasta: does environment matter? Chem Eng 58:859–864

Altamore L, Ingrassia M, Columba P, Chironi S, Bacarella S (2020) Italian consumers' preferences for pasta and consumption trends: tradition or innovation? J Int Food Agribus Market 32(4):337–360

Amicarelli V, Lagioia G, Bux C (2021) Global warming potential of food waste through the life cycle assessment: an analytical review. Environ Impact Assess Rev 91:106677

Arena U, Mastellone ML, Perugini F (2003) The environmental performance of alternative solid waste management options: a life cycle assessment study. J Chem Eng 96:207–222

Barton A, Beigg C, Macdonald I, Allison S (2000) High food wastage and low nutritional intakes among hospital patients. Clin Nutr 19(6):445–499

Beal T, Gardner CD, Herrero M, Iannotti LL, Merbold L, Nordhagen S, Mottet A (2023) Friend or foe? The role of animal-source foods in healthy and environmentally sustainable diets. J Nutr 153(2):409–425

Benvenga S, Famà F, Perdichizzi LG, Antonelli A, Brenta G, Vermiglio F, Moleti M (2022) Fish and the thyroid: a Janus Bifrons relationship caused by pollutants and the omega-3 polyunsaturated fatty acids. Front Endocrinol 13:891233

Benvenuti L, De Sants A, Ferrari M, Martone D, Rossi L (2022) The carbon footprint of Italian schools' meals: an optimal choice of dishes in vegan, vegetarian, and omnivorous menus. Front Nutr 31(9):854049

British Standard Institutions. Enabling everything with standards. Access date 20 June 2024. Available from: https://knowledge.bsigroup.com/

Brundtland GH (1987) Our common future: report of the world commission on environment and development. Geneva, UN-Dokument A/42/427

Bryan FL (1990) Hazard analysis critical control point (HACCP) systems for retail food and restaurant operations. J Food Prot 53(11):978–983

Bux C, Zizzo G, Amicarelli V (2023) A combined evaluation of energy efficiency, customer satisfaction and food waste in the healthcare sector by comparing cook-hold and cook-chill catering. J Clean Prod 429:139594

Carino S, Porter J, Malekpour S, Collins J (2020) Environmental sustainability of hospital foodservices across the food supply chain: a systematic review. J Acad Nutr Diet 120(5):825–873

Cerutti AK, Ardente F, Contu S, Donno D, Beccaro GL (2018) Modelling, assessing, and ranking public procurement options for a climate-friendly catering service. Int J Life Cycle Assess 23:95–115

Chiaverina P, Raynaud E, Fillâtre Ma, Nicklaus S, Bellassen V (2023) The drivers of the nutritional quality and carbon footprint of school menus in the Paris area. J Agric Food Ind Organ 21(2):147–169

De Boni A, Melucci FM, Acciani C, Roma R (2022) Community composting: a multidisciplinary evaluation of an inclusive, participative, and eco-friendly approach to biowaste management. Cleaner Environ Syst 6:100092

de Visser-Amundson A (2022) A multi-stakeholder partnership to fight food waste in the hospitality industry: a contribution to the United Nations Sustainable Development Goals 12 and 17. J Sustain Tour 30(10):2448–2475

Department of Health (2014) The Hospital Food Standards Panel's report on standards for food and drink in NHS hospitals. An independent group established by the Department of Health and led by Dianne Jeffrey, chairman of Age UK. Crown, United Kingdom

DOE Advanced Manufacturing Office (2022) Defining functional units for LCA and TEA. Access date 20 June 2024. Available from: https://www.energy.gov/sites/default/files/2022-06/2022-05-03%20-%20Functional%20Unit%20PDF_compliant.pdf

Donini LM, Castellaneta E, De Guglielmi S, De Felice MR, Savina C, Coletti C, Paolini M, Cannella C (2008) Improvement in the quality of the catering service of a rehabilitation hospital. Clin Nutr 27(1):105–114

Downer S, Berkowitz SA, Harlan TS, Olstad DL, Mozaffarian D (2020) Food is medicine: actions to integrate food and nutrition into healthcare. BMJ 29(369):2482

Driessen CE, Cameron AJ, Thornton LE, Lai SK, Barnett LM (2014) Effect of changes to the school food environment on eating behaviours and/or body weight in children: a systematic review. Obes Rev 15:968–982

Engelund EH, Lassen A, Mikkelsen BE (2007) The modernization of hospital foodservice—findings from a longitudinal study of technology trends in Danish hospitals. Nutr Food Sci 37:90–99

Ercin AE, Hoekstra AY (2012) Carbon and water footprints: concepts, methodologies and policy responses. United Nations Educational, Scientific and Cultural Organization, UNESCO

Eriksson M, Christensen J, Malefors C (2023) Making food waste illegal in Sweden—potential gains from enforcing best practice in the public catering sector. Sustain Prod Consumption 35:229–237

European Environment Agency. Composting. Access date 02 July 2024. Available from: https://www.eea.europa.eu/help/glossary/eea-glossary/composting

Eurostat. Glossary: incineration. Access date 02 July 2024. Available from: https://ec.europa.eu/eurostat/statistics-explained/index.php?title=Glossary:Incineration#:~:text=Incineration%20is%20a%20method%20of,for%20example%20for%20power%20generation

Fang K, Song S, Heijungs R, de Groot S, Dong L, Song J, Wiloso EI (2016) The footprint's fingerprint: on the classification of the footprint family. Curr Opin Environ Sustain 23:54–62

GBD 2019 Risk Factors Collaborators (2020) Global burden of 87 risk factors in 204 countries and territories, 1990–2019: a systematic analysis for the Global Burden of Disease Study 2019. Lancet 396:1223–1249

Global Burden of Disease Collaborative Network (2020) Global burden of disease study 2019. Results. Institute for Health Metrics and Evaluation. Access date 15 July 2024. Available from: https://vizhub.healthdata.org/gbd-results/

González N, Marquès M, Nadal M, Domingo JL (2020) Meat consumption: which are the current global risks? A review of recent (2010–2020) evidences. Food Res Int 137:109341

González-García S, Esteve-Llorens X, González-García R, González L, Feijoo G, Moreira MT, Leis R (2021) Environmental assessment of menus for toddlers serviced at nursery canteen following the Atlantic diet recommendations. Sci Total Environ 770:145342

Government Office for Science. Tackling obesities: future choices. Project Report, 2nd edn. Access date 15 July 2024. Available from: https://assets.publishing.service.gov.uk/media/5a759da7e5274a4368298a4f/07-1184x-tackling-obesities-future-choices-report.pdf

Grant F, Di Veroli JN, Rossi L (2023) Characterization of household food waste in Italy: three year comparative assessment and evaluation of seasonality effects. Waste Manag 164:171–180

Greenhouse Gas Protocol (2024) We set the standards to measure and manage emissions. Access date 20 June 2024. Available from: https://ghgprotocol.org

GURI (Gazzetta Ufficiale della Repubblica Italiana). Servizio di ristorazione e fornitura di derrate alimentare (approvato con DM n. 65 del 10 marzo 2020, in G.U. n. 90 del 4 aprile 2020). Access date 20 June 2024. Available from: https://www.mase.gov.it/sites/default/files/archivio/allegati/GPP/2021/cam_ristorazione.pdf

Hammer AJ, Millar C, Hennige SJ (2022) Reducing carbon emissions in aquaculture: using carbon disclosures to identify unbalanced mitigation strategies. Environ Impact Assess Rev 96:106816

Hebrok M, Boks C (2017) Household food waste: drivers and potential intervention points for design—an extensive review. J Clean Prod 151:380–392

Intergovernmental Panel on Climate Change. 2006 IPCC guidelines for national greenhouse gas inventories. Access date 20 June 2024. Available from: https://www.ipcc-nggip.iges.or.jp/public/2006gl/

International Panel for Climate Change (2006) IPCC guidelines for national greenhouse gas inventories. In: Eggleston HS, Buendia L, Miwa K, Ngara T, Tanabe K (eds) Institute for Global Environmental Strategies. Hyama, Japan.

International Panel for Climate Change (2021) IPCC sixth assessment report. Access date 20 June 2024. Available from: https://www.ipcc.ch/report/ar6/wg1/

ISO (2006a) ISO 14025:2006. Environmental labels and declarations—type III environmental declarations—principles and procedures. Access date 20 June 2024. Available from: https://www.iso.org/standard/38131.html
ISO (2006b) ISO 14044:2006. Environmental management—life cycle assessment—requirements and guidelines. Access date 20 June 2024. Available from: https://www.iso.org/standard/38498.html
ISO (2006c) ISO14040:2006. Environmental management—life cycle assessment—principles and framework. Access date 20 June 2024. Available from: https://www.iso.org/standard/37456.html
ISO (2018a) ISO 14064-1:2018. Greenhouse gases—Part 1: Specification with guidance at the organization level for quantification and reporting of greenhouse gas emissions and removal. Access date 20 June 2024. Available from: https://www.iso.org/standard/66453.html
ISO (2018b) ISO 14067:2018. Greenhouse gases—carbon footprint of products—requirements and guidelines for quantification. Access date 20 June 2024. Available from: https://www.iso.org/standard/71206.html
Jabeen F, Dhir A, Islam N, Talwar S, Papa A (2023) Emotions and food waste behavior: do habit and facilitating conditions matter? J Bus Res 155(Part B):113356
Khan S, Loyola C, Dettling J, Hester J, Moses R (2019) Comparative environmental LCA of the Impossible Burger® with conventional ground beef burger. Access date 02 July 2024. Available from: https://assets.ctfassets.net/hhv516v5f7sj/4exF7Ex74UoYku640WSF3t/cc213b148ee80fa2d8062e430012ec56/Impossible_foods_comparative_LCA.pdf
Kolbe K (2020) Mitigating climate change through diet choice: costs and CO_2 emissions of different cookery book-based dietary options in Germany. Adv Clim Chang Res 11:392–400. https://doi.org/10.1016/j.accre.2020.11.003
Laurent A, Owsianiak M (2017) Potentials and limitations of footprints for gauging environmental sustainability. Curr Opin Environ Sustain 25:20–27
Le TMH, Nguyen VKL, Le TTH, Nguyen TTH, Vu KN (2023) Customer satisfaction and fast-food restaurants: an empirical study on undergraduate students. J Foodserv Bus Res 26(6):771–792
Li J, Li W, Wang L, Jin B (2021) Environmental and cost impacts of food waste in university canteen from a life cycle perspective. Energies 14:5907
Mazzetto AM, Falconer S, Ledgard S (2023) Carbon footprint of New Zealand beef and sheep meat exported to different markets. Environ Impact Assess Rev 98:106946
Micha R, Karageorgou D, Bakogianni I, Trichia E, Whitsel LP, Story M, Penalvo JL, Mozaffarian D (2018) Effectiveness of school food environment policies on children's dietary behaviors: a systematic review and meta-analysis. PLoS ONE 13(3):e0194555
Ministero dell'Ambiente e della Sicurezza Energetica (2021) CAM—Criteri Ambientali Minimi. Access date 20 June 2024. Available from: https://gpp.mite.gov.it/Home/Cam#CamInVigore
Ministry of Health (2022) Linee di indirizzo nazionale per la ristorazione ospedaliera, assistenziale e scolastica. Access date 20 June 2024. Available from: https://www.salute.gov.it/imgs/C_17_pubblicazioni_3141_allegato.pdf
Mistretta M, Caputo P, Cellura M, Cusenza MA (2019) Energy and environmental life cycle assessment of an institutional catering service: an Italian case study. Sci Total Environ 657:1150–1160
Mondello G, Salomone R, Ioppolo G, Saija G, Sparacia S, Lucchetti MC (2017) Comparative LCA of alternative scenarios for waste treatment: the case of food waste production by the mass-retail sector. Sustainability 9:827
Mozaffarian D, Angell SY, Lang T, Rivera JA (2018) Role of government policy in nutrition: barriers to and opportunities for healthier eating. BMJ 361:2426
Murphy T (2017) The role of food in hospitals. HealthCareCAN. Leading. Innovation. Together. Ottawa, Canada
Official Journal of the European Union. Commission Delegated Decision (EU) 2019/1596 of 3 May 2019. Access date 20 June 2024. Available from: https://eur-lex.europa.eu/legal-content/EN/TXT/PDF/?uri=CELEX:32019D1597

OpenLCA Nexus (2024) Databases. Access date 20 June 2024. Available from: https://nexus.openlca.org/databases
Ormazabal M, Jaca C, Puga-Leal R (2014) Analysis and comparison of life cycle assessment and carbon footprint software. In: Xu J, Cruz-Machado V, Lev B, Nickel S (eds) Proceedings of the eighth international conference on management science and engineering management. Advances in intelligent systems and computing, vol 281. Springer, Berlin, Heidelberg
Paiva L, Hernández C, Santos D, Garayoa R, García L, Urdangarín C, Vitas AI (2022) Hospital plate waste assessment after modifications in specific dishes of flexible and inflexible food ordering systems. Sustainability 14:15616. https://doi.org/10.3390/su142315616
Pandey D, Agrawal M, Pandey JS (2021) Carbon footprint: current methods of estimation. Environ Monit Assess 178:135–160
Petersson T, Secondi L, Magnani A, Antonelli M, Dembska K, Valentini R, Varotto A, Castaldi S (2021) A multilevel carbon and water footprint dataset of food commodities. Sci Data 8:127
Piciocchi C, Lobefaro S, Luisi F, Miraglia L, Romito N, Luneia R, Foti S, Mocini E, Poggiogalle E, Lenzi A, Donini LM (2022) Innovative cooking techniques in a hospital food service: effects on the quality of hospital meals. Nutrition 93:111487
Pye-Smith C (2014) Scaling up the Brazilian school feeding model. Food and Agriculture Organization of the United Nations, Rome. Access date 15 July 2024. Available from: http://www.fao.org/3/a-h0050e.pdf
Rana RL, Bux C, Lombardi M (2023) Carbon footprint of the globe artichoke supply chain in Italy: from agricultural production to industrial processing. J Clean Prod 391:136240
Saget S, Porto Costa M, Sancho Santos C, Vasconcelos M, Styles D, Williams M (2021) Comparative life cycle assessment of plant and beef-based patties, including carbon opportunity costs. Sustain Prod Consumption 28:936–952
Sahal Alharbi N, Yahia Qattan M, Haji Alhaji J (2020) Towards sustainable food services in hospitals: expanding the concept of 'plate waste' to 'tray waste'. Sustainability 12:6872
Schanes K, Dobernig K, Gözet B (2018) Food waste matters—a systematic review of household food waste practices and their policy implications. J Clean Prod 182:978–991
Simon X, Copena D, Pérez-Neira D (2023) Assessment of the diet-environment-health-cost quadrilemma in public school canteens. An LCA case study in Galicia (Spain). Environ Dev Sustain 25:12543–12567
Tarnovskaya V (2023) Sustainability as the source of competitive advantage. How sustainable is it? In: Ghauri PN, Elg U, Hånell SM (eds) Creating a sustainable competitive position: ethical challenges for international firms. International business and management, vol 37. Emerald Publishing Limited, Leeds, pp 75–89
Torres-Escribano S, Vélez D, Montoro R (2010) Mercury and methylmercury bioaccessibility in swordfish. Food Addit Contam Part A Chem Anal Control Exposure Risk Assess Foreword 27(3):327–37
United States Environmental Protection Agency. Understanding global warming potential. Access date 20 June 2024. Available from: https://www.epa.gov/ghgemissions/understanding-global-warming-potentials
Vidal R, Moliner E, Pikula A, Mena-Nieto A, Ortega A (2015) Comparison of the carbon footprint of different patient diets in a Spanish hospital. J Health Serv Res Policy 20(1):39–44
Volanti M, Arfelli F, Neri E, Saliani A, Passarini F, Vassura I, Cristallo G (2022a) Environmental impact of meals: how big is the carbon footprint in the school canteens? Foods 11(2):193. https://doi.org/10.3390/foods11020193
Wackernagel M, Rees W (1996) Our ecological footprint: reducing human impact on the earth. New Society Publishers, Philadelphia, United States
Werthmann J, Renner F, Roefs A, Huibers MJH, Plumanns L, Krott N, Jansen A (2014) Looking at food in sad mood: do attention biases lead emotional eaters into overeating after a negative mood induction? Eat Behav 15(2):230–236
Wilfrid F, Peters MC (1980) Data sheet no. 160—Heat balances—boilers. In: Francis W, Peters MC (eds) Fuels and fuel technology, 2nd edn. Pergamon, pp 533–542

Willett W, Rockström J, Loken B, Springmann M, Lang T, Vermuelen S et al (2019) Food in the anthropocene: the EAT–lancet commission on healthy diets from sustainable food systems. Lancet 393(10170):447–492

Winther U, Ziegler F, Hognes ES, Emanuelsson A, Sund V, Ellingsen H (2009) Carbon footprint and energy use of Norwegian seafood products. SINTEF Fisheries and Aquaculture, Trondheim

World Health Organization (2021) Action framework for developing and implementing public food procurement and service policies for a healthy diet. World Health Organization, Geneva

World Health Organization (2020) Global health estimates 2019: deaths by cause, age, sex, by country and by region, 2000–2019. World Health Organization, Geneva

World Health Organization (2022a) Healthy diet. Access date 15 July 2024. Available from: https://www.who.int/news-room/fact-sheets/detail/healthy-diet

World Health Organization (2022b) Invisible numbers. The true extent of noncommunicable diseases and what to do about them. World Health Organization, Geneva

World Health Organization. Saving lives, spending less: the case for investing in noncommunicable diseases. Access date 15 July 2024. Available from: https://apps.who.int/iris/bitstream/handle/10665/272534/WHO-NMH-NVI-18.8-eng.pdf

Chapter 5
Sustainable Healthcare and Food Waste Minimization Through a Data-Driven Hospitality Management

Abstract Measurement and monitoring are fundamental activities that aim to effectively manage and mitigate the consequences of food waste. In Europe, food waste in foodservice is on average 10% across the entire supply chain, but there are no detailed data on food waste streams in hospitals. Food management in hospitals plays a significant role in the environmental, social, and economic health of patients and is intricately linked to malnutrition and patient recovery rates. The present chapter provides an overview of the most significant studies conducted in the domain of food waste management in hospitals. It discusses the implications for management and identifies potential strategies for improving sustainability in hospital foodservice. It examines trends, opportunities, and barriers in measuring and monitoring food waste, looking at patient satisfaction and nutrition as essential variables for sustainable healthcare systems, but also considering the carbon emissions of meals throughout their life cycle, from cradle to gate (including waste disposal) The chapter provides useful insights for setting environmentally friendly strategies and achieving sustainability excellence goals. It is aimed at academics and foodservice practitioners and provides useful insights for setting environmentally friendly strategies and achieving sustainability excellence goals.

Keywords Artificial intelligence · Data-driven hospitality · Digital technologies · Food waste measurement · Food waste monitoring

5.1 Introduction

Europe is paying increasing attention to food waste and is implementing specific statistical monitoring measures to assess the amount of food waste by sector, taking into account the NACE classification (i.e., primary production, processing and manufacturing, retail trade and other food distribution, restaurants, and food services) and by household (Eurostat 2024). The measurement and monitoring of food waste began in 2019 with the publication of Commission Delegated Decision (EU) 2019/1597 on a

C. Bux, *Strategies for a Sustainable and Environmentally Friendly Hospital Foodservice Management*, Environmental Humanities: Transformation, Governance, Ethics, Law 6,
https://doi.org/10.1007/978-3-031-82050-2_5

common methodology and minimum quality requirements for the uniform measurement of the amount of food waste, which supplements the Waste Framework Directive (2008/98/EC), that provided for an annual reporting obligation on the measurement of food waste by country (Official Journal of the European Union 2008). For 2021, it was estimated that 131 kg of food waste per inhabitant will be generated, of which around 10% will be generated in restaurants and catering. However, food waste in hospitals exceeds that of all other food supply sectors by a factor of 2 or 3 (Arriz-Jorquiera et al. 2024; Dias-Ferreira et al. 2015). The highest amount of food waste was recorded in Cyprus with 397 kg of food waste per capita per year, and the lowest in Slovenia with 68 kg.

At a global level, measuring food waste is still a complex challenge due to different measurement methods that can lead to different (and sometimes subjective) results (European Commission et al. 2022), and for this reason some countries are still unable to measure its quantity (Eurostat 2023). Specific challenges also relate to small sample sizes, incompleteness of sectoral surveys, suboptimal estimation of coefficients and misinterpretation of results, highlighting the need for a standardized definition, rules and common measurement methods. Comprehensive data on food waste is lacking worldwide. As proposed in Commission Delegated Decision (EU) 2019/1597, data can be collected using six different methods, namely direct measurement, waste composition analysis, mass balance, questionnaires and interviews, counting or scanning and diaries, with each method suitable for a single (or multiple) economic activity (see Chap. 1). In addition, the measurement of food waste can be carried out on the basis of proxy values and coefficients. Direct measurements or waste composition analysis, which are the more complex and expensive in terms of time and resources, are suitable for all economic activities, while diaries for restaurants and foodservice and counting/scanning are interesting methods. Counting/scanning, by definition the "assessment of the number of items that make up food waste and using the result to determine the mass" (Official Journal of the European Union 2019), requires a clear and comprehensive development of tools, also with recourse to artificial intelligence, and could represent one of the best solutions for the design of a data warehouse for food commodities or food waste.

According to the definition of minimum environmental criteria in Italy (Law 221/2015, supplemented by Legislative Decree 50/2016 and amended by Legislative Decree 56/2017), foodservice today is "the activity that includes the purchase of food and beverages, the preparation of meals with the food purchased, the transport and management of meals, the cleaning of the canteen, the premises of the cooking center and the equipment and dishes used, as well as the management of food surpluses resulting from the preparation and management of meals" (GURI 2020). This definition in terms of the life cycle concept (Baldwin et al. 2011) illustrates the complex nature of foodservice, which includes the delivery, preparation and transportation of food, but also the disposal of food waste, which differs from country to country (Stirnimann and Zizka 2022; McAdams et al. 2019) and can be reuse for human consumption or as animal feed, recycling, incineration with energy recovery or landfilling. In particular, the management of food and food waste represents a critical aspect in hospital catering, where food safety issues are intertwined with sustainable meals, nutritional/malnutrition rates and patient satisfaction for treatment and therapeutic purposes (Lins et al. 2021).

This chapter provides an overview of the most recent studies in the field of food waste measurement in hospital foodservice and discusses the most commonly used methodological approaches. Considering the strengths of measuring and monitoring food waste through counting/scanning methods, the study discusses the benefits of a comprehensive approach to measuring and monitoring food waste in hospital foodservice that considers patient satisfaction and nutrition as an essential variable for sustainable healthcare systems, but also the carbon emissions of meals along their entire life cycle, from cradle to gate (including waste management). The study provides researchers and practitioners with useful insights for the transition to real-time data in foodservice.

5.2 Materials and Methods

The study conducts a semi-systematic and critical literature review (Snyder 2019) on methods for measuring and monitoring food waste in hospital catering. Considering the variety of studies that have been conducted on food waste, i.e., studies on its economic, social or environmental aspects and impacts, the semi-systematic review (also known as "narrative") is suitable for examining food waste measurement methods, as reviews "often examine how research within a selected area has evolved over time" (Snyder 2019). In addition, this type of review helps to understand potential research traditions and synthesize findings using meta-narratives (Wong et al. 2013), and is important for identifying themes, theoretical perspectives, and common problems. The research conducts an explicit, transparent, and replicable analysis by applying the following step-by-step approach: (i) defining the objectives and scope of the research, including the timeframe, topic areas, document type, and language; (ii) identifying appropriate research strands; (iii) data collection and data synthesis; and (iv) data interpretation.

Based on the six measurement methods proposed in Commission Delegated Decision (EU) 2019/1597, the research investigates the measurement methods for food waste in hospital catering. The research was conducted on Scopus, which is considered a reliable collector of standardized, reputable and high-quality research, and considers a time frame of twenty years (2004–2024). The articles were searched using a TITLE-ABS-KEY approach. With regard to the type of document, a distinction was made between research and review articles, while book chapters or conference papers were not included in the search. In addition, only articles published in English were considered. In terms of search terms, the following combination of keywords was considered: "food waste" AND "hospital" OR "healthcare" to capture as many contributions as possible.

Metadata were catalogued in Microsoft Excel spreadsheets, indicating the type of research (i.e., research article, review article), publishing journal, measurement method, sample size, geographic area, and research focus, specifically environmental (i.e., amount of food waste, environmental impact of food waste), economic, or social (i.e., dietary rate). Specifically, papers were categorized into "quantitative studies" (i.e., studies that directly measure food waste with primary data) and "semi-quantitative studies" (i.e., studies that rely on secondary data or reviews), distinguishing between papers that directly measure and monitor food waste in hospitals, papers that review previous studies, and papers that address management practices for food waste management without quantitative measurement.

Of the 213 resulting studies from 1952 to 2024, only 25 met the objectives and scope of the research and therefore were selected for in-depth analysis. Table 5.1 illustrates the included contributions for review.

The bibliometric data was synthesized using VOSviewer, which examines the co-occurrence of keywords and shows the relationships and interactions between different research topics. VOSviewer is an open-source software that provides multi-dimensional scaling and clustering of research (Waltman et al. 2010) and helps to identify the evolution of research by creating suggestive maps of keywords and textual data through a rigorous method (Rana et al. 2022; van Eck and Waltman 2021).

5.3 Results

5.3.1 Bibliometric Results

VOSviewer software helps identify co-occurrences of keywords and highlights the relationship and the interaction among different research topics. It represents a suitable tool for identifying research trends, as well as a comprehensive overview of the research evolution, creating clear maps of keywords and textual data (Bux et al. 2022; Mulet-Forteza et al. 2019). Specifically, Fig. 5.1 illustrates the research trends identified using the VOSviewer software. To increase the reliability of the results, a minimum number of six keywords was considered in the research, and only 98 met the threshold (4.16%) out of 2355 keywords.

Figure 5.1 identifies three different research clusters, each of which focuses on a certain specific topic. Nonetheless, one can easily glimpse very close links between the different themes, highlighting the high transversality of the topic. It is therefore impossible (or almost) to do research on measuring and monitoring food waste in hospitals without having to consider the related environmental, economic and social aspects.

The first cluster, identified in red, includes 43 different keywords that relate to environment and waste management. Different concepts can be identified: on the one side, methods for measuring the environmental impacts; on the other side, the consequences of an unsustainable consumption and production pattern; last, the possible

Table 5.1 Selected studies for in-depth review

Type	Authors	Type	Journal	Area
Sqnt.	Hoteit et al. (2024)	Review	J. of Agriculture and Food Research	Generic
Qnt.	Arriz-Jorquiera et al. (2024)	Article	Waste Management	Florida, US
Qnt.	Jonathan et al. (2023)	Article	BMJ Open Quality	UK
Sqnt.	Diana et al. (2023)	Review	Nutrition & Food Science	Indonesia
Qnt.	Collins and Porter (2023)	Article	Nutrition & Dietetics	Australia
Sqnt.	Cook et al. (2023)	Review	Nutrition & Dietetics	Generic
Sqnt.	Antasouras et al. (2023)	Review	Int. J. Health Planning and Management	Generic
Sqnt.	Rinninella et al. (2023)	Review	Nutrients	Generic
Qnt.	Paiva et al. (2022)	Article	Sustainability	Spain
Qnt.	Berardy et al. (2022)	Article	Nutrients	California, US
Sqnt.	Cook et al. (2022)	Review	J. of Human Nutrition and Dietetics	Generic
Sqnt.	Porter and Collins (2021)	Article	J. Nutrition Education and Behavior	Australia
Qnt.	Thiel et al. (2021)	Article	PLoS One	New York, US
Sqnt.	Melikoglu (2020)	Article	Sustainable Production and Consumption	Turkey
Qnt.	Sahal Alharbi et al. (2020)	Article	Sustainability	Saudi Arabia
Qnt.	Gomes et al. (2020)	Article	Sustainability	Portugal
Sqnt.	Eriksson et al. (2020)	Article	Sustainability	Sweden
Qnt.	Bell et al. (2020)	Article	Appetite	Australia
Qnt.	Alshqaqeeq et al. (2020)	Article	Int. J. of Healthcare Management	Georgia, US
Q nt.	Schiavone et al. (2019)	Article	Int. J. Environ. Research and Public Health	Italy
Sqnt.	Jones (2019)	Article	British J. of Healthcare Management	UK
Sqnt.	Alshqaqeeq et al. (2018)	Review	Int. J. Healthcare Technology Management	Generic
Qnt.	Dias-Ferreira et al. (2015)	Article	Waste Management	Portugal
Qnt.	Sonnino and McWilliam (2011)	Article	Food Policy	UK

Note Qnt. = quantitative; Sqnt. = semi-quantitative; Qlt. = qualitative

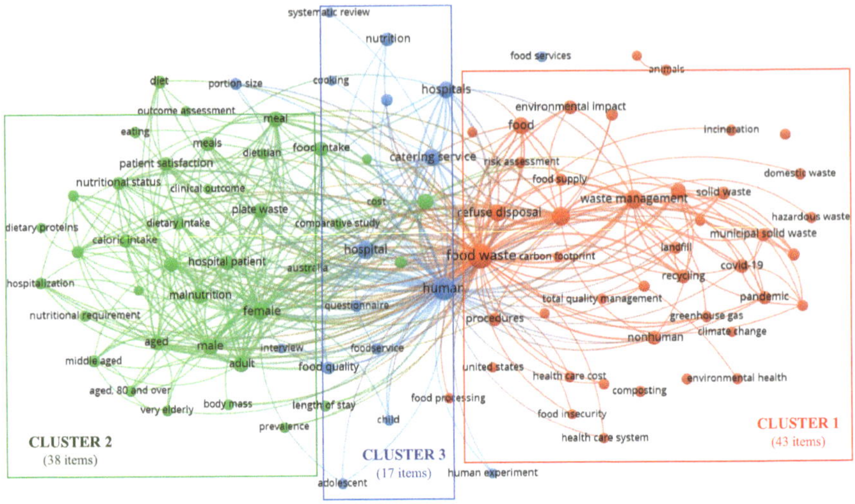

Fig. 5.1 Research trends and clusters through the VOSviewer software

solutions to reduce or valorize food waste. Specifically, in the field of methods for assessing the environmental impacts, the cluster identifies the "carbon footprint" or the "life cycle assessment", whereas in the field of the consequences it considers "climate change", "health costs" or "food insecurity". In terms of possible solutions, the cluster includes studies on the topics of "composting", "recycling", "incineration", or "landfill" but also protocols such as the "total quality management" (TQM). Such an approach is based on a twofold goal: quality of products, processes and systems, and customer satisfaction. Six are the main pillars for achieving the TQM, namely: (i) commitment; (ii) culture; (iii) continuous improvement; (iv) cooperation; (v) customer focus; and (vi) control (Psomas and Fotopoulos 2010). Last, the red cluster highlights a strong nexus between environmental impacts and environmental health, which is the branch of public health concerned with all aspects of the natural and anthropogenic environment that could affect human health.

The second cluster, identified with the green color, includes 38 keywords and related to consumer behavior and nutritional/dietary concerns. The main keywords included in the cluster are, for instance, "diet", "dietary intake", "energy intake", "dietary proteins", "nutritional assessment", "nutritional requirements", and "malnutrition", also considering users' characteristics like age or gender (i.e., "adult", "aged 80 and over", "male" and "female"), but also "length of hospitalization", which represents a significant variable in terms of healing. These topics, extremely interrelated, define the nexus between food intake, malnutrition (i.e., food waste) and opportunities of healing through sustainable and healthy consumption patterns.

Last, the third cluster identified in blue, includes 17 keywords and mainly relates to the methodological aspects used by researchers for carrying out their studies. Precisely, the cluster refers to keywords such as "human experiment", "interview", "quantitative analysis", "questionnaire", and "systematic review". Moreover, there are some keywords which highlight the importance of quantitative rather than qualitative studies, such as "portion size". Such element, highly stressed by several authors, could represent a possible key for reducing food waste among final users.

Considering the goal of the chapter, 25 contributions were selected. Out of 25 contributions, 13 were identified as "quantitative" and 16 as "semi-quantitative".

5.3.2 Findings from Quantitative Contributions

An essential part toward food waste reduction concerns its measurement: without a starting quantity it is not possible to predict its reduction over time. Table 5.2 presents the quantitative contributions selected for further review. These studies distinguish among food waste at breakfast, lunch and dinner and, at first glance, it appears that food waste quantities are rather similar across countries. Substantially, this also depends on the fact that hospital diets must provide similar nutritional values, regardless of the geographical area they belong to or culinary traditions. On average, food waste at breakfast is estimates at approximately 0.08 kg per patient per day, which upsurges at lunch (0.15–0.47 kg per patient per day) and dinner (0.11–0.48 kg per patient per day).

Based on the assumption that poor nutrition can lead to poorer outcomes for patients in hospitals, Jonathan et al. (2023) add a significant concept in food waste research, namely "portion sizes". From a methodological perspective, the study relies on primary data collected through pre- and post-meal weighting of meals in 21 patients on a single day, while qualitative information (e.g., reasons for not eating) was collected through face-to-face interviews. The results show that around 63–64% of food per patient is wasted at lunch, compared to 74–75% at dinner. There is an opportunity to reduce food waste by up to 4.41% by replacing large portion sizes (i.e., 0.22 kg of carbohydrates) with standardized portions (i.e., 170 kg of carbohydrates).

Collins and Porter (2023) conducted an audit of food waste (including food packaging waste) on 14 days of the week in Australia to determine the mass of food waste and the daily cost of sending it to landfill. Considering the daily diets, data were collected from Monday to Friday on the basis of 14 lunches and seven dinners, by distinguishing among different waste, as follows: (i) trayline waste, defined as the food leftovers after plating or food items still packaged but expired; (ii) plate waste, namely foods not consumed by patients and left into plates; and (iii) packaging waste, intending either plastic or paper containers from trays, single-use plastic dishes or other items required for serving foods. However, although the research considers different food waste categories and typologies, it does not account for preparation waste, which are quite complex to be accounted by staff members. In the field of hospital food waste, the purpose is to investigate customer satisfaction and malnutrition rates, whereas food waste at preparation does not affect this. However, in a more comprehensive and holistic research, these quantities should be accounted. In light of Collins and Porter (2023), food waste at dinner is higher that food waste at

Table 5.2 Quantitative findings of hospital food waste

Reference	Sample	Breakfast	Lunch	Dinner
Jonathan et al. (2023)	21 patients	N/D	63–64%	N/D
Collins and Porter (2023)	7 breakfast, 14 lunches, 7 dinners	35.2 kg per day	95.3 kg per day	97.2 kg per day
Paiva et al. (2022)	4641 meals (first phase)	0.08 kg per patient per day	0.22 kg per patient per day	0.23 kg per patient per day
	695 meals (second phase)	0.08 kg per patient per day	0.21 kg per patient per day	0.11 kg per patient per day
Berardy et al. (2022)	214 meat-containing meals	av. 0.29 kg per plate		
	233 vegetarian meals	av. 0.26 kg per plate		
Sahal Alharbi et al. (2020)	939 trays	0.08 kg per patient per day	0.15 kg per patient per day	0.16 kg per patient per day
Gomes et al. (2020)	321 meals	N/D	0.21 kg per patient per day	N/D
Alshqaqeeq et al. (2020)	2400 trays	av. 0.45 kg per meal per day		
Dias-Ferreira et al. (2015)	8000 meals	N/D	0.47 kg per patient per day	0.48 kg per patient per day
Sonnino and McWilliam (2011)	410 meals	av. 31%		

lunch, recording an amount of about 97.2 kg per day. In terms of trayline waste, it was estimated at 45.7 kg per day.

Food waste definitions may vary food waste records, as outlined in Chap. 1. Paiva et al. (2022) adopted the definition of plate waste and measured its amount through a direct measurement, namely directly weighing food waste after consumption. The study area is Spain. Specifically, the authors carried out a stepwise approach, distinguishing between a baseline scenario (without interventions) and an after-interventions scenario. In the first case, the sample was composed of 4641 meals, whereas in the second one of 695 meals. Likewise previous studies, dinners generate the highest amount of food waste (0.23 kg per patient per day), followed by lunch (0.22 kg per patient per day). Among different food items, fish represents the most wasted category (22–29%), then meat (about 24%), whereas the authors identified some opportunities for food waste reduction for milk (– 10%) and chicken (– 73%) soon after suitable education and awareness interventions.

Similarly to Paiva et al. (2022), also Berardy et al. (2022) adopted the plate waste definition and carried out an analysis on a seven-days period among 447 hospital patients in California. Specifically, the research distinguishes among plant-based (i.e., vegetable meals) and meat-based meals at breakfast, lunch and dinner. Paiva et al. (2022) recorded a higher percentage of food waste associated with meat-based meals, for an amount of about 0.29 kg per plate, whereas a slightly lower quantity in the field of plant-based ones, for an amount of 0.26 kg per plate. In terms of carbon emissions, as already discussed in Chap. 4, the authors identified a negative environmental impact due to meat-based meals, whose carbon emissions are quite 50% higher compared to plant-based ones. Specifically, one meat-based meal is responsible for about 0.60 kgCO_2eq, whereas a plant-based one for approximately 0.35 kgCO_2eq.

A more comprehensive research, which do not focus only on food waste in hospital, was carried out by Thiel et al. (2021). The purpose of this research was to quantify the amount of waste on a single weekday in a hospital in New York, distinguishing among solid trash, recyclables and compostable waste (both food and non-food but organic). Through a waste audit on trash bags, the authors carried out a double-step method, as follows: (i) first, they weighed every waste back leaving the kitchen during working hours; and (ii) secondly, they conducted in-depth audits on waste bags by verifying their specific content. As a result, Thiel et al. (2021) quantified an amount of 1.5 t of food waste per day, which means about 0.75 kg of food waste per meal (on a total of 2010 meals). In a long-run scenario, it was possible to predict an amount of 442 t of food waste per year, which is equal to about 294 tCO_2eq considering recycling for paper, metals and plastics, and all other waste sent to landfill. In other words, in light of Thiel et al. (2021), the carbon emissions associated with waste in hospitals are over 0.66 kgCO_2eq per kg of waste, including solid and organic components.

In 2020, Sahal Alharbi et al. (2020) sampled 939 meals in Saudi Arabia. From the methodological perspective, likewise Thiel et al. (2021), these authors adopted a double-step approach, as follows: (i) first, they gathered data related to tray components and weights of meals from the digital foodservice system, in terms of diets or food ingredients, for instance; and (ii) secondly, they measured food waste by conducting a daily weighing and sorting, which represents the most challenging stage, since it is expensive in terms of time, money and human resources. After sorting, it resulted that the entire amount of waste per patient per day is about 0.86 kg, distinguishing between: (a) 0.41 kg of food; (b) 0.30 kg of paper; (c) 0.13 kg of plastic; and (d) 0.02 kg of metals. In the field of food waste, Sahal Alharbi et al. (2020) also reported that isolated patients are more likely to waste less food (0.39 kg) compared to non-isolated ones (0.43 kg), and this could be due to the fact that many visiting relatives bring food from outside the canteen service, compromising the consumption of food provided by the hospital catering service.

Coming back to Europe, specifically in Portugal, Gomes et al. (2020) measured food waste among 321 hospital meals. Its originality lies in the analysis of different diets, i.e., standard pediatric, standard adult, light, soft texture and cream texture. To achieve this aim, the authors directly measured food waste by physically weighing meals before and after consumption. So as to guarantee reliability of results and reduce uncertainties, the measurement was conducted through high precision electronic kitchen scales. In percentage, pediatric patients tend to waste more food (about

72%) compared to oncology (47%) ones. In addition, female patients tend to waste more (61%) than male ones (50%). On average, it resulted that food waste per lunch was about 0.21 kg per patient per day, but clearly this varies greatly depending on the department and the type of patient, as demonstrated.

In Australia, Bell et al. (2020) conducted a research on the impacts of packaging on patients' dietary intake. Through this research, they could indirectly estimate food waste amounts among over 65-years old patients, since they carried out a direct measurement through electronic scale (i.e., CAS Smart Weighing Scale). In addition to direct measurement, Bell et al. (2020) introduced an interesting method, namely photographs of pre- and post-delivery trays. This method, which is based on 2D photographs, has now been partially overcome through the use of 3D scanners for the volumetric estimation of food waste. However, for its time, it represented a highly innovative and original instrument (2011).

Alshqaqeeq et al. (2020) investigated food waste in New York by weighing and recording data from approximately 2400 trays. On average, 940 meals per day generated 31.75 kg of solid food waste during meal preparation, 15.42 kg of liquid waste and 20.27 kg of food waste, corresponding to 0.03 kg, 0.01 kg and 0.02 kg per meal per day, respectively.

Schiavone et al. (2019) measured food waste in Italy using a questionnaire method. Specifically, the researchers asked a sample of 762 patients: "What percentage of your meals did you eat?" with reference to either the first plate, the second plate, the side plate or the fruit, and the possible answers were: (i) nothing/almost nothing; (ii) about ¼; (iii) about half; (iv) about ¾; and (iv) all/almost all. On average, the amount of food waste was 38.5% for the first plate, 39.7% for the second plate, 55% for the side plate, and 35.2% for the fruit.

One of the largest samples in the literature was achieved by Dias-Ferreira et al. (2015), namely 8000 meals over an 8-weeks period in Portugal. On average, each patient throws away 953 g of food each day, representing 35% of the food served, which equates 1.8 kgCO_2eq and 3.9 euro with acquisition and disposal of uneaten food per patient per day.

5.3.2.1 Quantitative Studies Based on Secondary Data

Arriz-Jorquiera et al. (2024) applied the mixed-integer programming models for single- and multi-objective optimization to minimize food waste and costs associated with meals. Data from a 1000-bed hospital in Florida, USA, was used to test the mathematical model. The percentage of food waste according to patient profile (e.g., hypertension, cardiovascular disease) was taken from the literature (Gomes et al. 2020). With regard to lunch meals, a food waste of 501.80 kg per week was determined in the baseline scenario, which means an average of 0.16 ($\pm$ 0.03 kg) kg per patient and lunch, with the possibility of reducing this to up to 0.13 kg per patient and lunch. The largest amount of food waste is generated on the pediatric ward (0.20 $\pm$ 0.03 kg per patient per lunch), followed by patients on a vegetarian diet

(0.19 ± 0.02 kg per patient per lunch). In terms of food groups, vegetables are the most wasted category (19 kg per day), followed by dairy products (17 kg per week).

In Sweden, Eriksson et al. (2020) estimated food waste reduction strategies through a questionnaire method, which was sent to 21 local authorities (at the regional scale). So as to confirm (or not) these qualitative results, responses were supplemented with food waste records from 20 hospitals. In terms of food waste amount, Eriksson et al. (2020) identified an average of 0.11 kg per patient per meal, distinguishing among different waste categories, i.e., plate waste (42%), serving waste (36%) and kitchen waste (22%). This difference illustrates that, of the total hospital food waste, a part is the responsibility of the patient, while another is associated with food preparation operations and is therefore inevitable (i.e., unavoidable food waste). This study, although it uses methodologies for quantifying food waste compatible with the provisions of the reference legislation (Commission Delegated Decision 2019/1597 of May 3, 2019), openly declares that measurement is an essential, but not sufficient, element for reducing food waste. It highlights the need for control measures and incentives (i.e., economic incentives). To this extent, different strategies could be implemented to reduce food waste along the entire supply chain, from preparation to consumption, and are based on three pillars, as follows: (i) prevention, which means avoid food from becoming surplus food; (ii) human reuse, which means avoid surplus food from becoming food waste; and (iii) animal reuse, material recycling, nutrient recovery or energy recovery, which means avoid food waste from becoming food loss (Lombardi and Costantino 2021). In this case, the definitions of surplus food, food waste and food loss are new and are reinterpreted from a sustainability perspective (Teigiserova et al. 2020), namely: (i) "food waste is defined as surplus food that is not reused for feeding people" (Garrone et al. 2014); and (ii) "food loss only refers to the streams that are truly lost, whether because not accounted for or disappearing from the accounting" (Teigiserova et al. 2020). Among other, some of the main strategies to reduce food waste in the upstream stages are food redistribution activities, which include "various types of organizations and initiatives that distribute edible food that is about to be wasted, directly or indirectly to food insecure people" (Poore and Nemececk 2019; Vittuari et al. 2017), but also awareness campaign or food-sharing platforms (if possible). From a consumer behavior perspective, Porter and Collins (2021) investigated perceptions, beliefs and expectations among 40 hospital patients in four hospital in Australia. Although economic costs represent one of the main reasons for reducing food waste (and one of the main levers for sustainability at households), this rule does not apply to hospitals, where the cost of the meal is not directly borne by the patient/consumer. The main variables to consider at hospital level are essentially food/appetite, food quality and quantity, as well as the foodservice model. This highlights the main role in the reduction of food waste by companies that deal with foodservice, noting the individual behavior of the patient/consumer in a minimal way. In other words, if the food is of good quality, then it will be consumed more. But good quality, of course, depends on the company that provides the foodservice.

Melikoglu (2020) created a semi-empirical model in Turkey to estimate energy consumption, food waste, and the energy contents connected with it using data from each bed and unit area. Based on Turkey's annual food waste generation of roughly 49 kt, it was determined that these quantities could produce between 0.87 and 0.97% of the energy utilized in hospitals. Based on quantitative research, the Eastern Mediterranean Region's hospitals squander an average of 25.4% of their food, whereas Indonesia's hospital waste is higher at 27.66%. Notwithstanding the fact that the primary objectives of reviews are to quantify and measure food waste, their contributions primarily take a localized approach, failing to consider global issues.

5.3.3 *Findings from Semi-quantitative Contributions*

Semi-quantitative studies identify contributions that measure and monitor food waste on the basis of secondary data or reviews, which means that these researches did not carry out a primary measurement of food waste. In particular, these studies tried to fill in data gaps by relying on other studies or by using mathematical models, so as to integrate either environmental, nutritional or economic variables and develop a comprehensive assessment.

Table 5.3 illustrates the main findings from previous qualitative studies.

Hoteit et al. (2024) reviews previous studies on food waste in the Mediterranean Region, precisely the eastern countries of the region. The purpose of the research is

Table 5.3 Findings from semi-quantitative contributions

Reference	Findings
Diana et al. (2023)	21 studies included in the review. The median plate waste is 27.6%, from 14.8 to 88.7%. Food waste costs from 0.07 to 0.5 USD per patient per day
Rinninella et al. (2023)	19 studies included in the review. Strategies to reduce food waste regard protected mealtimes and volunteer feeding assistance, food presentation strategies, nutritional counseling and education and plant-based proteins meal
Cook et al. (2023)	Four studies and 81 grey literature in the review. Strategies to reduce food waste are composting, donation of surplus or animal feeding. Benefits are the reduction of hauling feed, staff handling, and waste sent to landfill
Antasouras et al. (2023)	Strategies to minimize food waste are flexible portion sizes, increased food choices through selective menus, additional nutritional support and an enhanced ordering and delivery system
Cook et al. (2022)	Eight studies and nine gray literature in the review. Common practices are two-week data collection, foodservice staff data collection, direct measurement of plate and tray waste, use of electronic scales
Alshqaqeeq et al. (2018)	92 studies included in the review. One of the best options to reduce food waste regards tailoring the food choice of individual patients

to identify research gaps related to quantification of food waste, as well as to define possible managerial strategies toward its reduction. In the research, which reviews the main contributions published in recent years, six studies were includes, as follows: (i) Al-shoshan (1992), who conducted a study in Saudi Arabia among 554 patients and 205 attendants on a two-days period; (ii) Dehnavi et al. (2017), who studied food residuals among 425 patients admitted to general surgery, cardiac, internal medicine, thoracic, pediatric, cerebrovascular, oncology, burns, skin, poisoning, and accidents wards, over one-week; (iii) Simzari et al. (2017), who conducted a research among 120 patients with an hospitalization time of more than seven days in three different educational university hospitals in Iran; (iv) Sahal Alharbi et al. (2020), who investigated food waste from 939 trays over a three-week period; (v) Al-Muhannadi et al. (2021), who carried out a food waste analysis among 563 patients on a period of six months, from January to June; and (vi) Anari et al. (2024), who investigated food waste among 12 patients, nine nurses, three hospital managers, three food suppliers, eight nutritionists and 13 members of the kitchen staff.

One of the main insights was highlighted by Jones (2019), which declared that hospital foodservice still relies on low-tech measurement methods, such as regular or electronic weighing scale. The use of direct measurement tools are highly expensive in terms of time, economic resources and personnel to be involved in the study. Furthermore, although scales can be precise and reliable, human intervention is always necessary, making the measurement process sometimes random or partially discretionary (e.g., tiredness, inattention). For this reason, over time, precision instruments have been developed for measuring food waste.

Nowadays there are high-tech methods to measure food waste at foodservice (e.g., workplace canteens, restaurants, schools), for examples those developed by Winnow (https://www.winnowsolutions.com) or Nuvilab (https://www.nuvilab.com) that include the use of artificial intelligence and scanning tools for the identification of food groups and the related costs or carbon emissions. These tools, which are based both on the use of scanners for food recognition and on scales for quantifying food waste, represent the most developed technologies worldwide. The patents, which will be analyzed in Sect. 5.4.1, take into account the consequences of food waste in environmental, economic and social terms.

In terms of artificial intelligence to reduce food waste, Ani and Boateng (2024) discussed the nexus between artificial intelligence and food waste reduction by investigating the potential of the Winnow software (so-called Winnow Waste Monitor). It is based on the use of a camera and a smart scale connected to a data analysis and visualization system, and its utility regards the determination of food that are thrown away by kitchen staff in restaurants. This tool provides daily, weekly and monthly reports that encompass both economic and environmental implications of food waste (Principato et al. 2023). However, although artificial intelligence could represent an essential tool to reduce food waste, its potential has not yet been adopted in hospital foodservice and canteens. Moreover, reports and data should always be implemented by human intelligence, since managers and staff must discuss priorities and find solutions together (i.e., briefing) and take action by integrating changes based on common decision (Principato et al. 2023).

5.4 Theoretical and Managerial Implications

Measurement and monitoring tools are required to enhance the sustainability of hospital foodservice by reducing food waste and their associated environmental (i.e., carbon emissions), social (i.e., nutrition rate) and economic (i.e., waste management costs) implications. Environmentally friendly benefits and competitivity in foodservice should be based on a comprehensive concept of "data-driven sustainability", according to different steps, namely: (i) definition of industry-specific goals; (ii) development of a theoretical framework; and (iii) combination of human and artificial intelligence.

Opportunities to adopt standardized methods should leverage on artificial intelligence technologies, which can accurately track food consumption patterns, identify food waste quantities and composition and provide, if possible, areas of intervention, without forgetting the importance of human expertise. Specifically, food waste measurement should be based on real-time monitoring systems that continuously analyzes food consumption and waste, by relying on digital scales or scanners.

Nowadays, there are some organizations that adopt proper waste audits (Bux et al. 2023), but it is limited in time, for instance one single weekday, or one or two weeks, or to a relatively small sample size, which does not allow for a continuous replicability and representativeness of the analysis over time. Latest studies estimated food waste among 21 patients (Jonathan et al. 2023), whereas the largest sample size refers to about 8000 meals, but the research was cost-effective and was conducted approximately ten years ago (Dias-Ferreira et al. 2015). Moreover, the majority of studies that rely on primary data tent to estimate only the mass of food waste, without considering additional environmental or economic implications.

Berardy et al. (2022), Thiel et al. (2021) and Dias-Ferreira et al. (2015) included a carbon footprint analysis of food waste. Specifically, Berardy et al. (2022) estimated the environmental impacts associated with food waste by combining SimaPro life cycle assessment software and previous studies to fill in eventual data gaps, but the study considers a cradle-to-farm approach without including disposal pathways (e.g., recycling, landfilling, etc.) emissions. Furthermore, Thiel et al. (2021) evaluated the greenhouse gas emissions associated with disposal and treatment of waste, by distinguishing a recycling, composting or landfilling scenario, focusing on the sole waste management phase. Last, Dias-Ferreira et al. (2015) estimated the carbon emissions by considering that 1 t of food waste generates about 1.9 tCO_2eq, but it does not distinguish neither per food category nor per waste management pathway. However, it lacks a systematic assessment of the emissions associated with food waste, which considers the impacts of food production (from cradle-to-gate), meals preparation (from gate-to-gate) and waste disposal (from gate-to-grave). Carbon emissions must consider either the emissions associated with material input (i.e., agricultural processed food, animal husbandry, fishing and crops), energy input (i.e., foodservice type, distinguishing between cook-hold or cook-chill, for instance), or the material and energy output (e.g., landfilling, anaerobic digestion, composting).

Practically, comprehensive studies should be based on the assumption that food waste impacts twice, namely when it is produced and when it is disposed, and such a double track is still underrated from the environmental and the economic side. In addition, studies should consider that food waste impacts also on the social side, specifically in the field of patients' nutritional intake (or rate of malnutrition). Currently, it seems that no study focuses on the nutritional impacts of food waste on the individual tray and on the individual patient, not considering that such an analysis, on the contrary, could contribute to improving the patient's well-being and the possibility of recovery. Unlike other foodservice activities, hospital catering takes on a prominent role in patient care, and it is responsible and indispensable for supporting human health.

The role of measuring and monitoring food waste is essential to make hospital catering services efficient and effective in every respect. Considering the importance of guaranteeing nutritious and quality food, monitoring food waste (systematically and in real-time) would allow researchers to collect useful data and carry out economic, environmental, and nutritional assessments. Specifically, healthcare personnel responsible for the care of patients with metabolic diseases, but also cancer patients, the elderly and children (but not only), through the quantification of food waste, could measure nutrient losses and calculate, if necessary, the rate of malnutrition associated with each patient. This leads to the need to create tools based on a data-driven approach and artificial intelligence (segmentation, for example). Foodservice companies must necessarily comply with food traceability rules to ensure food safety and compliance with mandatory regulations at the local and the international level. Therefore, companies must necessarily collect data and can use it independently. From this, companies can use this food data to support the needs of healthcare workers and hospital users; therefore, the monitoring of food waste can effectively become a tool for improving healthcare performance and, in a broader sense, hospital management itself.

5.4.1 Food Waste Measurement and Management Through Artificial Intelligence Technologies

Nowadays, the nexus between artificial intelligence and environmental strategies has assumed a significant role, particularly with regard to technologies that can be used to simplify data collection and provide useful information to researcher, practitioners and public authorities (Elouariaghli et al. 2022; de Jesus et al. 2021). In the context of food waste measurement and monitoring, digital technologies may help researchers collect and analyze data, as well as solve problems instead of human beings. However, these algorithms must be tested to increase their reliability, since their efficiency is not yet 100% accurate (de Jesus et al. 2021; Pishgar-Komleh et al. 2020). The need to develop artificial intelligence tools to collect data is due to the "time-consuming", "tedious" or "impossible to practice" (Ibn-Mohammed et al. 2023; Akhshik et al. 2022) process of data collection according to current methods (e.g., direct measurement, waste composition analysis). These activities can be partially resolved by introducing artificial intelligence tools such as machine learning, visual computing, mathematical artificial neutral networks and adaptive neuro-fuzzy inference system methods.

By definition, artificial intelligence is the "science and engineering of making intelligent machines, especially intelligent computer programs" and its tasks are addressed "to understand human intelligence without confining to methods that are biologically observable" (McCarthy 2024). Currently, one-quarter of organizations are relying on artificial intelligence tools to address labor shortages, but not only: its use has spread in the field of customer service, computer vision and automated stock trading (IBM 2024), but the possibilities for development are endless, even in foodservice. Historically, the first question related to artificial intelligence was: "Can machine think?" by Alain Turin in 1950 (Turing 1950), whereas its first definition was introduced during the Dartmouth Conference in 1956 (Delipetrev et al. 2020). It first use was addressed to solve algebraic problems, language translations and geometric theorems, and its information technology infrastructure was rather simple.

Onyeaka et al. (2023) defined the main artificial intelligence tools for addressing food waste toward sustainable goals. Specifically, it identifies: (i) machine learning, which is suitable for analyzing consumer behavior patterns to predict food purchase, reduce overproduction or optimize crop yields on the basis of the weather patterns and the soil conditions, for instance (Van Klompenburg et al. 2020; Garre et al. 2020); (ii) image recognition, which can be used in quality control for food items and could avoid food waste by identifying substandard products before reaching customers (Sundaram and Zeid 2023); (iii) natural language processing, that can interpret feedbacks from customers to reduce food waste; (iv) Internet of Things (IoT), which is essential to collect data about food storage conditions in order to prevent spoilage or to improve shelf-life (Popa et al. 2019); (v) blockchain, which can ensure traceability and decrease food waste and frauds; and (vi) reinforcement learning, that can optimize food logistics and supply chain management (Bačiuliene et al. 2023).

In the field of foodservice, artificial intelligence software were developed by different companies, such as Winnow, Kitro or Leanpath, which operate in restaurants, hotels, cafeterias, but also in hospital or school canteens. In brief, these companies rely on 3D cameras electric scales and digital platforms (i.e., dashboards for presenting results), but behind the "surface" they adopt machine learning algorithms, visual computing, sensor technologies and data analytics tools.

Leanpath was developed in 2004 and distinguishes between kitchen waste and plate waste, according to the phase in which it occurs. In terms of kitchen waste, it is based on three pillars, namely: (i) track food waste; (ii) discover food waste prevention opportunities; and (iii) drive results with a series of prevention tools. Specifically, in terms of food waste tracking, Leanpath offers different measurement tools, such as the Leanpath 360 and 360-AI series 3 or the Leanpath 360 Floor Scale Series 3, which include a proper digital scale, or the Leanpath Station, the Leanpath Tablet or the Leanpath Go, which do not include a digital scale. These tools, especially those based on artificial intelligence technologies, capture images of food waste and after a short period of training they can recognize food waste (in terms of food typology). However, all of them are based on electronic or digital scales for measuring food waste (Leanpath 2024).

More recently, Winnow was developed in 2013. It offers different solutions, such as: (i) the VisionControl, which is based on an image led system with an in-built camera and a connected scale; (ii) the VisionAI, which offer more accurate data and minimal efforts, and includes also a food cost profile and an automated recording process; and the (iii) VisionAI+, which offers a lifetime monthly coaching session and provides the highest quality data through a validation of a data specialist. Either Leanpath or Winnow offer analytics (through a clear dashboard for practitioners) and action planning solutions by using historic waste data to predict trends and by proposing smart reduction tips (Winnow 2024). Precisely, Winnow identified economic and environmental costs associated with food waste, but it is based on a digital scale for measuring food waste, lacking a proper volumetric assessment for identifying food waste quantities only through a 3D camera.

Last, Kitro technologies were introduced in 2017, and is rather similar to the previous technologies (Kitro 2024).

5.4.1.1 Previous Studies on the Use of Artificial Intelligence for Food Waste Measurement

The subject matter of food waste is extensive, and the reference literature is replete with diverse perspectives, case studies, and analyses that examine its economic, social, and environmental ramifications. Several literature reviews were dedicated to the topic, especially quantitative and qualitative aspects.

In the quantitative field, the authors focused on the measurement methodologies (Amicarelli and Bux 2021) or the environmental impacts associated with the different valuation pathways (Wang et al. 2024). Specifically, researchers examined alternative strategies for the valorization of food waste, examining the challenges and opportunities associated with anaerobic digestion, incineration, or hydrothermal carbonization (Shukla et al. 2024), and introducing the concept of "upcycling" to mitigate climate change and resource depletion (Dou et al. 2024; Rakesh and Mahendran 2024).

From a qualitative perspective, the literature has emphasized the drivers and obstacles to the prevention or reduction of food waste (Chia et al. 2024), with distinctions made according to the stage of the supply chain. For example, Lagioia et al. (2024) identified digitalization as the key to minimizing waste. Samiun et al. (2024) noted that only 193 healthcare food-waste documents were published in the past 70 years, with the oldest dating back to 1952. Furthermore, despite the small number of contributions on the topic, it was shown that there was an annual growth of 24.55% from 2011 to 2022, confirming the increasing interest in food waste studies.

However, when it comes to food waste measurement using primary data and the use of digital technologies for its reduction, the area remains relatively unexplored. One of the first to develop a combination of Bayesian network and agent-based modeling to classify household food waste drivers, but no mention is made to its quantification (Anggraeni et al. 2021). Then, Onyeaka et al. (2023) presented some digital technologies to tackle food waste toward a circular economy, such as machine learning, artificial intelligence image recognition, natural language processing, the Internet of Things (IoT), blockchain, and reinforcement learning. A thorough examination has led to the identification of several solutions, including smart real-time monitoring and analysis, predictive analytics, smart inventory management, consumer awareness and education through AI-powered apps, and donation and distribution. A significant point in the field of digital technologies was highlighted by Trevisan and Formentini (2023). Foodservice is the less considered investigated by researchers (i.e., out of 93 different studies along the entire agri-food chain, only four focus on foodservice) IoT, blockchain, digital twins, intelligence/active packaging or robotic process automation are some of the technologies that could be used to reduce food waste, but only AI can do that.

In the healthcare sector, these monitoring needs through digital technologies are multiplied by the different interests at stake: economic, for the reduction of costs linked to the supply of resources and waste disposal; social, to ensure equity among patients, increase consumer satisfaction and minimize the level of malnutrition.

5.5 Conclusions

The chapter, through a semi-systematic and critical literature review, explored the methods used to measure food waste in hospital foodservice by focusing on the role of artificial intelligence to enhance (and simply) data collection. In light of the Commission Delegated Decision 1597/2019, which identified six methods to measure food waste (i.e., direct measurement, waste composition analysis, mass balance, questionnaire and interviews, counting or scanning and diaries), the chapter highlighted that breakfast food waste is about 0.08 kg per patient per day, whereas lunch food waste is 0.15–0.47 kg per patient per day, with average values between 0.21 and 0.22 kg. Moreover, dinner food waste ranges from 0.11 to 0.48 kg per patient per day, with relatively lower values compared to lunch.

Nowadays, the nexus between artificial and human intelligence must be discussed in the realm of food waste. Specifically, foodservice should adopt digital tools (and not only conventional ones) to measure and monitor food waste and reduce its related environmental, economic and social consequences. The use of conventional scales must be supplemented by digital ones or innovative instruments that are based on photographs or 3D scanners, to reduce human efforts and guarantee reliability of results (e.g., Leanpath, Winnow, Kitro). Furthermore, the use of artificial intelligence that can segment food waste could return results of considerable importance, especially in hospitals, where details on the food waste quantities and the relative micro and macronutrients loss is essential.

References

Akhshik M, Bilton A, Tjong J, Veer Singh C, Faruk O, Sain M (2022) Prediction of greenhouse gas emissions reductions via machine learning algorithms: toward an artificial intelligence-based life cycle assessment for automotive lightweighting. Sustain Mater Technol 31:e00370

Al-Muhannadi A, Vishwas S, Estoya S, Hamatosa J, Daradkeh G (2021) Assessment and evaluation of food waste at Al-Khor Hospital—Hamad Medical Corporation in the state of Qatar. Food Sci Qual Manag 109:25–32

Al-shoshan AA (1992) Study of the regular diet of selected hospitals of the ministry of health in Saudi Arabia: edible plate waste and its monetary value. J R Soc Health 112(1):7–11

Alshqaqeeq F, Twomey JM, Overcash MR (2018) Food waste in hospitals: review. Int J Healthc Technol Manag 17(2–3):186–196

Alshqaqeeq F, Twomey JM, Overcash M, Sadkhi A (2020) A study of food waste in St. Francis hospital. Int J Healthc Manag 13:24–32

Amicarelli V, Bux C (2021) Food waste measurement toward a fair, healthy and environmental-friendly food system: a critical review. Br Food J 123(8):2907–2935

Anari R, Nikooyeh B, Ghodsi D, Amini M, Neyestani TR (2024) An in-depth analysis of hospital food waste in terms of magnitude, nutritional value, and environmental and financial perspectives: a cross-sectional study. Waste Manag Res J Sustain Circular Econ 42(2):167–177

Anggraeni MC, Silaban CA, Anggreainy MS, Cahyadi E (2021) Role of artificial intelligence in the management of food waste. In: 2nd International conference on artificial intelligence and data sciences (AiDAS). IPOH, Malaysia, pp 1–6

Ani OG, Boateng ID (2024) Improving health through sustainable and healthy urban food system policy in Nigeria. Dietetics 3:1–15

Antasouras G, Vasios GK, Kontogiorgis C, Ioannou Z, Poulios E, Deligiannidou GE, Troumbis AY, Giaginis C (2023) How to improve food waste management in hospitals through focussing on the four most common measures for reducing plate waste. Int J Health Plann Manag 38(2):296–316

Arriz-Jorquiera M, Acuna JA, Rodríguez-Carbó M, Zayas-Castro JL (2024) Hospital food management: a multi-objective approach to reduce waste and costs. Waste Manag 175:12–21

Bačiuliene V, Bilan Y, Navickas V, Lubomír C (2023) The aspects of artificial intelligence in different phases of the food value and supply chain. Foods 12:1654

Baldwin C, Wilberforce N, Kapur A (2011) Restaurant and food service life cycle assessment and development of a sustainability standard. Int J Life Cycle Assess 16:40–49

Bell AF, Tapsell LC, Walton K, Batterham M (2020) Unwrapping nutrition: exploring the impact of hospital food and beverage packaging on plate waste/intake in older people. Appetite 144:104463

Berardy A, Egan B, Birchfield N, Sabaté J, Lynch H (2022) Comparison of plate waste between vegetarian and meat-containing meals in a hospital setting: environmental and nutritional considerations. Nutrients 14:1174

Bux C, Aluculesei AC, Moagăr-Poladian S (2022) How to monitor the transition to sustainable food services and lodging accommodation activities: a bibliometric approach. Sustainability 14(15):9102

Bux C, Zizzo G, Amicarelli V (2023) A combined evaluation of energy efficiency, customer satisfaction and food waste in the healthcare sector by comparing cook-hold and cook-chill catering. J Clean Prod 429:139594

Chia D, Yap CC, Wu SL, Berezina E, Aroua MK, Gew LT (2024) A systematic review of country-specific drivers and barriers to household food waste reduction and prevention. Waste Manag Res 42(6):459–475

Collins J, Porter J (2023) Quantifying waste and its costs in hospital foodservices. Nutr Diet 80(2):192–200

Cook N, Collins J, Goodwin D, Porter J (2022) A systematic review of food waste audit methods in hospital foodservices: development of a consensus pathway food waste audit tool. J Hum Nutr Diet 35:68–80

Cook N, Goodwin D, Porter J, Collins J (2023) Food and food-related waste management strategies in hospital food services: a systematic review. Nutr Diet 80(2):116–142

de Jesus JO, Oliveira-Esquerre K, Medeiros DL (2021) Integration of artificial intelligence and life cycle assessment methods. IOP Conf Ser Mater Sci Eng 1196:012028

Dehnavi Z, Faghani M, Khorasanchi Z, Safari Bidokhti Z, Seifi N, Norouzy A (2017) Investigation of the volume of food waste in Qaem and Imam Reza Hospitals in Mashhad, Iran. J Nutr Fasting Health 5(4):178–183

Delipetrev B, Tsinarakii C, Kostic U (2020) Historical evolution of artificial intelligence. EUR 30221EN, Publications Office of the European Union, Luxembourg, JRC120469

Diana R, Martianto D, Baliwati YF, Sukandar D, Hendriadi A (2023) Food waste in Indonesian hospitals: a systematic review. Nutr Food Sci 53(5):81–900

Dias-Ferreira C, Santos T, Oliveira V (2015) Hospital food waste and environmental and economic indicators—a Portuguese case study. Waste Manag 46:146–154

Dou Z, Dierenfeld ES, Wang X, Chen X, Shurson GC (2024) A critical analysis of challenges and opportunities for upcycling food waste to animal feed to reduce climate and resource burdens. Resour Conserv Recycl 203:107418

Elouariaghli N, Kozderka M, Quaranta G, Pena D, Rose B, Hoarau Y (2022) Eco-design and life cycle management: consequential life cycle assessment, artificial intelligence and green IT. IFAC-PapersOnLine 55(5):49–53

Eriksson M, Malefors C, Bergström P, Eriksson E, Persson Osowski C (2020) Quantities and quantification methodologies of food waste in Swedish Hospitals. Sustainability 12:3116

European Commission (2022) Guidance on reporting of data on food waste and food waste prevention according to Commission Implementing Decision (EU) 2019/2000. Directorate E: Sectoral and Regional Statistics

Eurostat (2023) Food waste and food waste prevention—estimates. Access date 03 July 2024. Available from: https://ec.europa.eu/eurostat/statistics-explained/index.php?title=Main_Page

Eurostat. Food waste and food waste prevention by NACE Rev. 2 activity—tonnes of fresh mass. Access date 03 July 2024. Available from: https://ec.europa.eu/eurostat/databrowser/view/env_wasfw/default/table?lang=en

Garre A, Ruiz MC, Hontoria E (2020) Application of Machine Learning to support production planning of a food industry in the context of waste generation under uncertainty. Oper Res Perspect 7:100147

Garrone P, Melacini M, Perego A (2014) Surplus food recovery and donation in Italy: the upstream process. Br Food J 116:1460–1477

Gomes A, Saraiva C, Esteves A, Gonçalves C (2020) Evaluation of hospital food waste—a case study in Portugal. Sustainability 12:6157

GURI (Gazzetta Ufficiale della Repubblica Italiana). Servizio di ristorazione e fornitura di derrate alimentare (approvato con DM n. 65 del 10 marzo 2020, in G.U. n. 90 del 4 aprile 2020). Access date 03 July 2024. Available from: https://www.mase.gov.it/sites/default/files/archivio/allegati/GPP/2021/cam_ristorazione.pdf

Hoteit M, Antar E, Malli D, Fattouh F, Khattar M, Baderddine N, El Cheikh A, Abiad MG, Hassan HF (2024) A review on hospital food waste quantification, management and assessment strategies in the eastern Mediterranean region. J Agric Food Res 15:100959

IBM. What is artificial intelligence (AI)? Access date 17 July 2024. Available from: https://www.ibm.com/topics/artificial-intelligence

Ibn-Mohammed T, Mustapha KB, Abdulkareem M, Ucles Fuensanta A, Pecunia V, Dancer CEJ (2023) Toward artificial intelligence and machine learning-enabled frameworks for improved predictions of lifecycle environmental impacts of functional materials and devices. MRS Commun 13:795–811

Jonathan A, Reid K, Radcliffe T (2023) Managing food waste in the inpatient population. BMJ Open Qual 12(4):002436

Jones H (2019) Managing food waste across hospitals sites and trusts. Br J Healthc Manag 25(5)

Kitro. Our solution. Access date 17 July 2024. Available from: https://www.kitro.ch/product

Lagioia G, Amicarelli V, Strippoli R, Bux C, Gallucci T (2024) Sustainable and circular practices in the hotel industry in Southern Italy: opportunities, barriers and trends in food waste management. Br Food J 126(1):428–452
Leanpath. Solutions. Access date 17 July 2024. Available from: https://www.leanpath.com/solutions/
Lins M, Puppin Zandonadi R, Raposo A, Ginani VC (2021) Food waste on foodservice: an overview through the perspective of sustainable dimensions. Foods 10(6):1175
Lombardi M, Costantino MA (2021) Hierarchical pyramid for food waste based on a social innovation perspective. Sustainability 13:4661
McAdams B, von Massow M, Gallant M, Hayhoe MA (2019) A cross industry evaluation of food waste in restaurants. J Foodserv Bus Res 22(5):449–466
McCarthy J. What is artificial intelligence? Access date 17 July 2024. Available from: https://www-formal.stanford.edu/jmc/whatisai.pdf
Melikoglu M (2020) Appraising food waste generation and forecasting food waste to energy potentials of hospitals in Turkey: a global to local analysis. Sustain Prod Consumption 24:292–297
Mulet-Forteza C, Genovart-Balaguer J, Mauleon-Mendeza E, Merigó JM (2019) A bibliometric research in the tourism, leisure and hospitality fields. J Bus Res 101:819–827
Official Journal of the European Union. Commission delegated decision (EU) 2019/1596 of 3 May 2019. Access date 03 July 2024. Available from: https://eur-lex.europa.eu/legal-content/EN/TXT/PDF/?uri=CELEX:32019D1597
Official Journal of the European Union. Directive 2008/98/EC of the European Parliament and of the Council of 19 November 2008 on waste and repealing certain Directives. Access date 03 July 2024. Available from: https://eur-lex.europa.eu/legal-content/EN/TXT/PDF/?uri=CELEX:02008L0098-20180705
Onyeaka H, Tamasiga P, Nwauzoma UM, Miri T, Juliet UC, Nwaiwu O, Akinsemolu AA (2023) Using artificial intelligence to tackle food waste and enhance the circular economy: maximising resource efficiency and minimising environmental impact: a review. Sustainability 15:10482
Paiva L, Hernández C, Santos D, Garayoa R, García L, Urdangarín C, Vitas AI (2022) Hospital plate waste assessment after modifications in specific dishes of flexible and inflexible food ordering systems. Sustainability 14:15616
Pishgar-Komleh SH, Akram A, Keyhani A, Sefeedpari P, Shine P, Brandao M (2020) Integration of life cycle assessment, artificial neural networks, and metaheuristic optimization algorithms for optimization of tomato-based cropping systems in Iran. Int J Life Cycle Assess 25:620–632
Poore J, Nemecek T (2019) Reducing food's environmental impacts through producers and consumers. Science 360:987–992
Popa A, Hnatiuc M, Paun M, Geman O, Hemanth DJ, Dorcea D, Son LH, Ghita S (2019) An intelligent iot-based food quality monitoring approach using low-cost sensors. Symmetry 11:374
Porter J, Collins J (2021) A qualitative study exploring hospital food waste from the patient perspective. J Nutr Educ Behav 53(5):410–417
Principato L, Marchetti S, Barbanera M, Ruini L, Capoccia L, Comis C, Secondi L (2023) Introducing digital tools for sustainable food supply management: tackling food loss and waste in industrial canteens. J Ind Ecol 27:1060–1075
Psomas EL, Fotopoulos CV (2010) Total quality management practices and results in food companies. Int J Product Perform Manag 59(7):668–687
Rakesh B, Mahendran R (2024) Upcycling of food waste and food loss—a sustainable approach in the food sector. Trends Food Sci Technol 143:104274
Rana RL, Bux C, Lombardi M (2022) Trends in scientific literature on the environmental sustainability of the artichoke (*Cynara cardunculus* L. spp.) supply chain. Br Food J 125(6):2315–2332
Rinninella E, Raoul P, Maccauro V, Cintoni M, Cambieri A, Fiore A, Zega M, Gasbarrini A, Mele MC (2023) Hospital services to improve nutritional intake and reduce food waste: a systematic review. Nutrients 15(2):310

Sahal Alharbi N, Yahia Qattan M, Haji Alhaji J (2020) Towards sustainable food services in hospitals: expanding the concept of 'plate waste' to 'tray waste'. Sustainability 12:6872

Samiun NAA, Razalli NH, Shahar S, Manaf ZA, Kefeli Z, Khudzari JM, Jamhuri N (2024) The global research trends on food waste at healthcare setting: a bibliometric analysis using Scopus database. Malays J Med Health Sci 20(1):312–322

Schiavone S, Pelullo CP, Attena F (2019) Patient evaluation of food waste in three hospitals in Southern Italy. Int J Environ Res Public Health 16:4330

Shukla KA, Sofian ADABA, Singh A, Chen WH, Show PL, Chan YJ (2024) Food waste management and sustainable waste to energy: Current efforts, anaerobic digestion, incinerator and hydrothermal carbonization with a focus in Malaysia. J Cleaner Prod 448:141457

Simzari K, Vahabzadeh D, Saeidlou S, Khoshbin S, Bektas Y (2017) Food intake, plate waste and its association with malnutrition in hospitalized patients. Nutr Hosp 34:1376–1381

Snyder H (2019) Literature review as a research methodology: an overview and guidelines. J Bus Res 104:333–339

Sonnino R, McWilliam S (2011) Food waste, catering practices and public procurement: a case study of hospital food systems in Wales. Food Policy 36(6):823–829

Stirnimann A, Zizka L (2022) Waste not, want not: managerial attitudes towards mitigating food waste in the Swiss-German restaurant industry. J Foodserv Bus Res 25(3):302–328

Sundaram S, Zeid A (2023) Artificial intelligence-based smart quality inspection for manufacturing. Micromachines 14:570

Teigiserova DA, Hamelin L, Thomsen M (2020) Towards transparent valorization of food surplus, waste and loss: clarifying definitions, food waste hierarchy, and role in the circular economy. Sci Total Environ 706:136033

Thiel CL, Park S, Musicus AA, Agins J, Gan J, Held J, Horrocks A, Bragg MA (2021) Waste generation and carbon emissions of a hospital kitchen in the US: potential for waste diversion and carbon reductions. PLoS ONE 16(3):0247616

Trevisan C, Formentini M (2023) Digital technologies for food loss and waste prevention and reduction in agri-food supply chains: a systematic literature review and research agenda. IEEE Trans Eng Manag

Turing AM (1950) Computing machinery and intelligence. The essential Turing: the ideas that gave birth to the computer age. Jack Copeland B (ed). Oxford UP, Oxford, pp 433–464

van Eck NJ, Waltman L (2021) VOSviewer manual. Version 1.6.17. Universiteit Leiden, Leiden

Van Klompenburg T, Kassahun A, Catal C (2020) Crop yield prediction using machine learning: a systematic literature review. Comput Electron Agric 177:05709

Vittuari M, De Menna F, Gaiani S, Falasconi L, Politano A, Dietershagen J, Segrè A (2017) The second life of food: an assessment of the social impact of food redistribution activities in Emilia Romagna, Italy. Sustainability 9:1817

Waltman L, van Eck NJ, Noyons ECM (2010) A unified approach to mapping and clustering of bibliometric networks. J Infometrics 4(4):629–635

Wang H, Luo N, Liu Y, Gong Y (2024) Evaluating the environmental impact of agri-food waste reduction: a bibliographic analysis and conceptual framework. Environ Impact Assess Rev 105:107425

Winnow. Solutions. Access date 17 July 2024. Available from: https://www.winnowsolutions.com/product/food-waste-management-software

Wong G, Greenhalgh T, Westhorp G, Buckingham J, Pawson R (2013) RAMESES publication standards: meta-narrative reviews. BMC Med 11:20

Chapter 6
Prospects for a Sustainable Foodservice Management

Abstract Foodservice plays an essential role in the world economy, especially when it comes to hospital, welfare, and social-healthcare facilities. A healthy and correct diet is important for both environmental protection and for improving the health conditions of patients. Catering organizations must adopt a life cycle approach, considering both food supply and waste disposal. The final chapter of this book summarizes the entire study conducted, and the analyses presented, outlines some research directions and future perspectives, with a view to energy efficiency, circular economy, and, more generally, sustainable development. Moreover, the chapter provides some guidelines for practitioners to integrate the theoretical concepts of a solely academic nature. In future, it is required a closer collaboration between universities, research bodies, and organizations in order to share know-how, expertise, and conduct experiments applied to organizational realities. Environmental protection, economic growth (also through the reduction of costs associated with energy or waste disposal), and the enhancement of life expectancy (through improved clinical and socio-health practices) are among the most significant factors in achieving the desired sustainable development.

Keywords Artificial intelligence · Circular economy · Energy efficiency · Minimum environmental criteria · Sustainable development

6.1 Introduction

In conclusion, the book entitled "Strategies for a sustainable and environmentally friendly hospital foodservice management" aimed to furnish the fundamental components necessary to comprehend the significance of foodservice on the economic, environmental, and social levels. Despite its significant and profitable economic impact, foodservice plays a pivotal role in social development by providing essential services. Consider schools, universities, or work canteens, as well as hospital canteens. But also consider the quantity of meals each day and the associated environmental impacts. In order to ensure that everything works correctly (and sustainably, as has been repeated

C. Bux, *Strategies for a Sustainable and Environmentally Friendly Hospital Foodservice Management*, Environmental Humanities: Transformation, Governance, Ethics, Law 6,
https://doi.org/10.1007/978-3-031-82050-2_6

several times over the course of the five previous chapters), operators in the sector must consider energy efficiency, waste disposal, the quality of food being processed of supply, as well as the level of satisfaction of the end customer (i.e., the patient).

One last look at the numbers. In the year 2021, the market size of the global foodservice industry was estimated to be approximately 2.52 trillion USD, with an anticipated rise to 4.43 trillion USD by the year 2028 (Statista 2024a). The catering sector in Europe had an annual turnover of approximately 473 billion USD, representing 18% of the global amount. There are over 8.32 million employees in food and beverage activities, with Germany having the highest number of employees (1.67 million), followed by Spain (1.27 million), and Italy (1.26 million) (Statista 2024b). However, from an environmental perspective, several negative impacts are associated with the entire lifecycle of the foodservice, from natural resource (e.g., energy, water) consumption during the upstream and core stages until the waste (and food waste) generation during the downstream stages (Meier et al. 2021; Mistretta et al. 2019). Food production, consumption, and waste are responsible for having negative environmental impacts on the environment. Food waste represents approximately 6% of the total greenhouse gas emissions on a global scale (Krstić et al. 2024; Luthin et al. 2023). The impact associated with land use change, deforestation and organic soil management is 4.4 $GtCO_2eq$. When examining single food commodities, it was determined that beef is responsible for the highest quota of emissions, with an estimated amount of 99.48 $kgCO_2eq$ per kilogram of product, followed by lamb with 39.72 $kgCO_2eq$, and beef for dairy production with 33.3 kgCO2eq. On the contrary, plant-based products are accountable for relatively low amounts of carbon emissions, such as maize (1.7 gCO_2eq), wheat and rye (1.57 $kgCO_2eq$), and potatoes (0.46 $kgCO_2eq$) (Poore and Memecek 2018; Ritchie et al. 2022).

Sector operators are encouraged to collaborate with research bodies and universities to identify holistic strategies that take into account environmental, economic, and social aspects. However, the current challenges are becoming increasingly complex, necessitating the requirement of specialized skills, research and development, digital technologies, innovation, and artificial intelligence. Quality is no longer sufficient to look at as compliance with market standards, but as a set of transversal aspects that range from respect for the environment to subjective taste, but which also take into account the local context, the aesthetics of the foods, the personnel service, and much more.

Therefore, a question arises: What is the future of collective catering services?

6.2 The Future of Hospital Foodservice

6.2.1 *A Summary of the CAMs*

One of the first challenges of hospital foodservice concerns compliance with the "minimum environmental criteria for the foodservice and foodstuff" (trad. from the Italian: "Criteri ambientali minimi per il servizio di ristorazione collettiva e fornitura di derrate alimentari", herein CAMs), introduced by the Ministerial Decree no. 65 of 9 May 2020 in Italy (GURI 2020). Without compliance, it would be almost impossible to participate in tenders and start your own economic activity. In hospitals, it is necessary to differentiate menus according to age groups and pathologies, which increases the complexity of supply chains and customer satisfaction management. In particular, it is imperative to distinguish between hospital departments for neonatology, pediatrics, and other users aged 0–19, and patients in departments other than pediatrics, non-pediatric patients in care, detention, and similar facilities (retirement homes).

In hospitals, foods such as fruit and vegetables, cereals (including pasta, rice, etc.), legumes and eggs, as well as meat and fruit baby food, and jams and marmalades, must be organic. For other types of food, such as meat, fish, cured meats, cheeses, milk (and derivatives), oil, peeled tomatoes, and fruit juices, the same guidelines identified for school canteens apply, which are more stringent and require specific attention (GURI 2020). But it is also important to consider all the rules relating to the preparation of the food. Therefore, the challenges do not solely pertain to supply, but also encompass proficiency in the preparation of meals. When meals containing both organic and conventional ingredients are prepared in the same cooking facility, it is imperative to guarantee the separation of organic and conventional ingredients at all stages of storage and processing. Therefore, if the preparation of meals with organic products cannot take place in dedicated areas or lines, separation can be guaranteed on a temporal basis, i.e., preparing meals at different times. This requires having well-organized premises for activities and staff who are qualified and competent (GURI 2020). In terms of transparency, the presence of organic and otherwise qualified foods must be communicated to patients, their family members, or other assistant staff.

Considering that the rules for the upstream and core phases of foodservice are rather strict and leave no room for discretion, the issue of waste management becomes more complex. Although there exist regulations for the management and disposal of surplus food (and food waste), what if one wishes to resort to preventive measures? According to the CAMs, food waste is recorded at 35–40% of the number of meals produced in catering schools and around 30% in hospitals (GURI 2020). In particular, significant volumes of waste are recorded, of the order of 26.7% and 30.1% in standard diets and common food. In 18.8% of the meals investigated, the waste was greater than 50% (GURI 2020; Ministero della Salute 2024). In general, the calories discarded in the hospital meals provided to patients constitute 27.8% of the total. This causes a continuous increase in patient malnutrition during the hospital stay, which further exacerbates the patient's condition and increases hospitalization times by 30–40% (GURI 2020). Although Chap. 4 presents percentages, on average, slightly lower (i.e., food waste at lunch is approximately 8.6–13.9%, whereas at dinner it increases up to 18%, with the highest peak at dinner at about 30.7%), it still represents one of the major challenges for sustainability and nutrition.

The rules identified by the CAMs for the management of food waste are simple and leave room for operator discretion regarding methodologies and tools. Precisely, food surpluses must be calculated, at least approximate, and monitored, distinguishing between first and second courses, side dishes, fruit, and main courses. Every six months, investigations must be carried out through specific questionnaires or interviews, to analyze food satisfaction and comprehend the drivers of food waste. Moreover, without prejudice to specific needs based on pathologies or particular clinical needs, changes to the menus and weights must be proposed, based on the results of screening on the nutritional status of patients and the detection of quantities of food consumed.

6.2.2 The Role of Artificial Intelligence in the Measurement of Food Waste

There are diverse perspectives, case studies and analyses that examine the economic, social and environmental ramifications of food waste in the reference literature (Wang et al. 2024; Amicarelli and Bux 2020), focusing on both quantitative (Caldeira et al. 2019; Corrado et al. 2019) and qualitative aspects (Hebrok and Boks 2017; Thyberg and Tonjes 2016). Today, the nexus between artificial intelligence and environmental strategies has assumed a significant role, particularly with regard to technologies that can be used to simplify data collection and provide useful information to researcher, practitioners and public authorities (Elouariaghli et al. 2022; de Jesus et al. 2021). Data collection and analysis, as well as resolving issues, may be aided by digital tools in the realm of food waste measurement and surveillance. Precisely, Onyeaka et al. (2023) presented some digital technologies to tackle food waste toward a circular economy, such as machine learning, artificial intelligence image recognition, natural language processing, the Internet of Things (IoT), blockchain, and reinforcement learning. Moreover, there are several solutions that have been identified, including smart real-time monitoring and analysis, predictive analytics, smart inventory management, consumer awareness and education through artificial intelligence-powered apps, and donation and distribution. From an academic and bibliometric perspective, Trevisan and Formentini (2023) highlighted that only four out of 93 different studies along the entire agri-food chain focus on foodservice, which is the least considered investigated by researchers. However, even though IoT, blockchain, digital twins, intelligence/active packaging or robotic process automation could be used to cut down on food waste throughout the entire supply chain, only artificial intelligence can accomplish that in the food industry.

Some technologies already exist for measuring food waste at the foodservice level (e.g., Winnow, Kitro or Leanpath), but future research will need to focus on segmentation technologies (Aslan et al., 2020) that recognize, more rapidly and more immediately, the quantity and type of wasted food. In conclusion, what are the next steps? First, it is essential to rely on a 3D intelligent camera (Tran and Lin, 2019). The trays returned after hospital consumption in the cooking centers will need to be photographed. Visual computing and machine learning methods will be used to train AI to create a specialized database of the collective catering service, based on both standard and custom diets that are given to patients every day. A daily dashboard will be constructed by identifying the qualitative (type of food) and quantitative (weight of food waste) characteristics using segmentation techniques. The food trend, including consumption and waste, will be monitored over time and for each individual patient if possible. Ultimately, the information gleaned through the use of AI will be crucial in identifying tactics for minimizing waste (to lessen environmental and financial burdens) and boosting patient nutrient intake.

Future steps for measuring food waste and monitoring its quantities over time are summarized in Fig. 6.1.

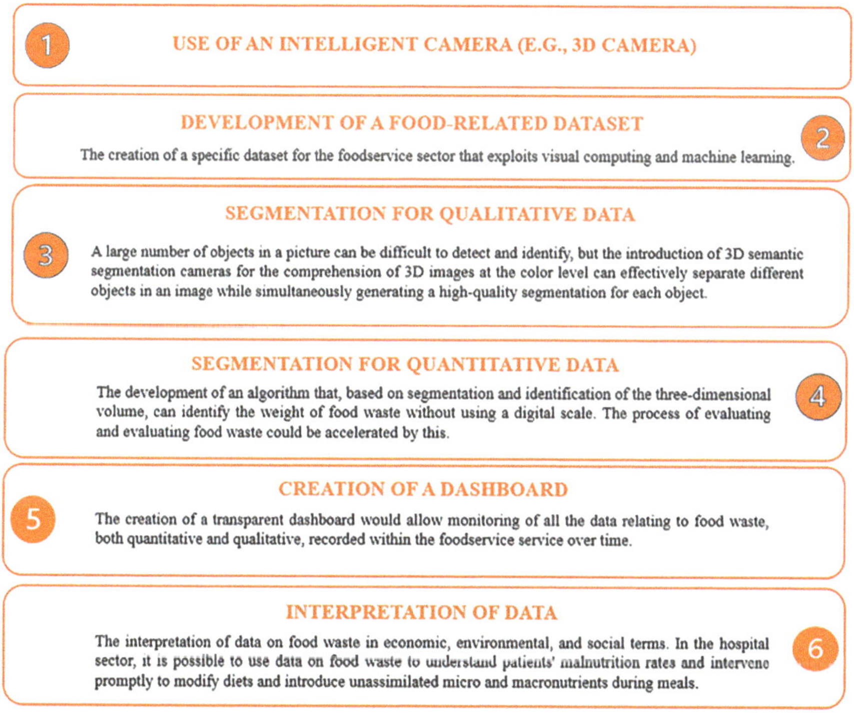

Fig. 6.1 Steps for measuring food waste through digital technologies and artificial intelligence

6.3 Conclusions

The transition from a conventional foodservice to a sustainable one is already in progress, and both companies and public authorities have demonstrated that they are in line with sustainability objectives and consumer needs. The empirical evidence presented by the case studies in this volume has demonstrated that the reduction of food waste and its associated emissions of greenhouse gases can be achieved through meticulous and analytical management of catering services, with particular consideration given to the requirements of hospital patients. The main challenges relate to a daily monitoring, so that each patient (with their pathologies, hospitalization times, personal tastes, etc.) can achieve the highest level of satisfaction. Thus, the concluding equation is quite straightforward: the greater the client contentment, the higher his nutrient intake and the greater his therapeutic possibilities, the greater the reduction in food wastage and the consequent financial and environmental burdens. Therefore, it is essential to implement customer satisfaction monitoring systems through digital technologies and artificial intelligence.

References

Amicarelli V, Bux C (2020) Food waste measurement toward a fair, healthy and environmental-friendly food system: a critical review. Br Food J 123(8):2907–2935

Aslan S, Ciocca G, Mazzini D, Schettini, R. Benchmarking algorithms for food localization and semantic segmentation. Int J Mach Learn Cybern 11:2827–2847

Caldeira C, De Laurentiis V, Corrado S, van Holsteijn F, Sala S (2019) Quantification of food waste per product group along the food supply chain in the European Union: a mass flow analysis. Resour Conserv Recycl 149:479–488

Corrado S, Caldeira C, Eriksson M, Hanssen OJ, Hauser HE, van Holsteijn F, Liu G, Oestergren K, Parry A, Secondi L, Stenmarck A, Sala S (2019) Food waste accounting methodologies: challenges, opportunities, and further advancements. Glob Food Sec 20:93–100

de Jesus JO, Oliveira-Esquerre K, Medeiros DL (2021) Integration of artificial intelligence and life cycle assessment methods. IOP Conf Ser Mater Sci Eng 1196:012028

Elouariaghli N, Kozderka M, Quaranta G, Pena D, Rose B, Hoarau Y (2022) Eco-design and life cycle management: consequential life cycle assessment, artificial intelligence and green IT. IFAC-PapersOnLine 55(5):49–53

GURI (Gazzetta Ufficiale della Repubblica Italiana). Servizio di ristorazione e fornitura di derrate alimentare (approvato con DM n. 65 del 10 marzo 2020, in G.U. n. 90 del 4 aprile 2020). Access date 23 July 2024. Available from: https://www.mase.gov.it/sites/default/files/archivio/allegati/GPP/2021/cam_ristorazione.pdf

Hebrok M, Boks C (2017) Household food waste: drivers and potential intervention points for design—an extensive review. J Clean Prod 151:380–392

Krstić M, Elia V, Agnusdei GP, De Leo F, Tadić S, Miglietta PP (2024) Evaluation of the agri-food supply chain risks: the circular economy context. Br Food J 126(1):113–133. https://doi.org/10.1108/BFJ-12-2022-1116

Luthin A, Traverso M, Crawford RH (2023) Circular life cycle sustainability assessment: an integrated framework. J Ind Ecol 28:41–58

Meier T, von Borstel T, Welte B, Hogan B, Finn SM, Bonaventura M, Friedrich S, Weber K, Dräger de Teran T (2021) Food waste in healthcare, business and hospitality catering: composition,

environmental impacts and reduction potential on company and national levels. Sustainability 13:3288
Ministero della Salute. Linee di Indirizzo Nazionale per la Ristorazione Ospedaliera ed Assistenziale. Access date 23 July 2024. Available from: https://www.aslvc.piemonte.it/images/downloads/prevenz/Docs_SIAN/linee_guida_nazionali_ristorazione_ospedaliera.pdf
Mistretta M, Caputo P, Cellura M, Cusenza MA (2019) Energy and environmental life cycle assessment of an institutional catering service: an Italian case study. Sci Total Environ 657:1150–1160
Onyeaka H, Tamasiga P, Nwauzoma UM, Miri T, Juliet UC, Nwaiwu O, Akinsemolu AA (2023) Using artificial intelligence to tackle food waste and enhance the circular economy: maximising resource efficiency and minimising environmental impact: a review. Sustainability 15:10482
Poore J, Nemecek T (2018) Reducing food's environmental impacts through producers and consumers. Science 360(6392):987–992
Ritchie H, Rosado P, Roser M (2022) Environmental impacts of food production. Access date 17 June 2024. Available from: https://ourworldindata.org/environmental-impacts-of-food
Statista (2024a) Net turnover of the food and beverage service activities industry in the European Union (EU-27) in 2021 and 2022 (in million euros). Access date 17 June 2024. Available from: https://www.statista.com/statistics/1459198/net-turnover-food-beverage-service-activities-industry-eu/
Statista (2024b) Number of employees in the food and beverage service activities industry in Europe in 2022, by country. Access date 17 June 2024. Available from: https://www.statista.com/statistics/1459209/number-of-employees-food-beverage-services-europe-by-country/
Thyberg KL, Tonjes DJ (2016) Drivers of food waste and their implications for sustainable policy development. Resour Conserv Recycl 106:110–123
Tran VL, Lin HY (2019) Development of a 3D semantic segmentation camera based on mask regional convolutional neural network. In: 2019 International conference on system science and engineering (ICSSE), Dong Hoi, Vietnam, pp 98–1020
Trevisan C, Formentini M (2023) Digital technologies for food loss and waste prevention and reduction in agri-food supply chains: a systematic literature review and research agenda. IEEE Trans Eng Manag
Wang H, Luo N, Liu Y, Gong Y (2024) Evaluating the environmental impact of agri-food waste reduction: a bibliographic analysis and conceptual framework. Environ Impact Assess Rev 105:107425

Summary

The book "Strategies for a Sustainable and Environmentally Friendly Hospital Food-service Management" presents both theoretical and practical content, and is intended for researchers, academics, and practitioners. Its primary objective is to contextualize the phenomenon of sustainability in a holistic and academic-organizational manner. To create synergies among all stakeholders of the foodservice sector, the topic of sustainability and the reduction of food waste must be treated from a theoretical, practical, technical, and corporate perspective.

One of the main challenges relating to the reduction of food waste is linked to the different definitions of the phenomenon and the non-harmonized and non-standardized methodologies for its measurement. It is evident from literature that the diverse methodologies for evaluating food waste proposed by the Delegated Decision of the European Commission 2019/1597 are suitable for one or more phases of the production chain. Therefore, a major process of conceptualization and knowledge of the supply chain is required, as well as the definition of the system boundaries for the measurement and monitoring phase. In this regard, it is imperative to promote certain measurement methodologies, such as material flow analysis and counting/scanning techniques. However, these methodologies necessitate direct measurement, which currently entails significant expenditures in terms of time and human resources. This leads to the need to implement digital technologies and artificial intelligence to speed up (and mechanize) the material measurement process, leaving the interpretation of the results and the definition of strategies for reducing food waste to human intervention.

In Italy, through the introduction of minimum environmental criteria (CAM) with Ministerial Decree no. 65 of 9 May 2020, the essential role of school, hospital and commercial (e.g., offices, barracks, etc.) foodservice operators was highlighted for the pursuit of sustainability objectives and for the reduction of the environmental impacts associated with production and food disposal. The reduction of environmental impacts associated with the production of goods can be achieved through

C. Bux, *Strategies for a Sustainable and Environmentally Friendly Hospital Foodservice Management*, Environmental Humanities: Transformation, Governance, Ethics, Law 6, https://doi.org/10.1007/978-3-031-82050-2

the supply of organic food, especially fruit and vegetables. Furthermore, an overarching objective is the reduction of animal-based diets and the use of low animal protein diets, especially for hospitals, offices, barracks, and universities. Furthermore, sustainability strategies include the use of reusable tableware, the abandonment of pre-processed foods or fifth range foods, or the abandonment of single dose foods. Last, the supply of local foods is of particular importance, with the dual objective of reducing the environmental impacts associated with the transport of goods and increasing the turnover of small and medium-sized local agricultural businesses.

However, environmental sustainability in catering services also involves energy efficiency, with reference to logistics, transport and waste disposal services. This volume, through an empirical case conducted in nine hospitals and 60 departments on a sample of 1590 meals per day for lunch and dinner, evaluates energy consumption and possible efficiency strategies by comparing cook-and-chill and cook-and-hold food production.

As a result, the cook-and-hold method is more efficient than the cook-and-chill method, with a 42% reduction in diesel consumption and a 93% reduction in electricity consumption, since the latter relies on many fewer steps and energy-intensive operations. It eliminates the blast freezing and cold storage phase at very low temperatures. Furthermore, it replaces transport with refrigerated trucks with that with non-refrigerated trucks and reduces the temperature of the trolleys for service by approximately 50% (from 120 °C to approximately 65 °C). These changes generally do not compromise the quality of the service (consumer preference is even higher), therefore being preferable from an environmental point of view. However, many hospital facilities continue to require in tender notices that the service be provided in cook-and-chill, raising hypotheses of contamination and lack of food safety in the case of the cook-and-hold method.

A further empirical study, based on a sample of 3454 patients over three years of analysis, highlighted some possibilities for improving sustainability performance in the foodservice sector through a carbon footprint analysis. Considering hospital diets, which by law must respect specific characteristics, it was demonstrated that it is useful to modify diets with reference to some foods. Dairy products should be avoided, as they have a very high environmental impact and are not particularly appreciated by consumers/patients. Beef meat could be replaced, with the same nutritional values, with chicken meat and boned pork meat, recording a reduction in carbon footprint of approximately 84% and 77%, respectively. Ham should be replaced with boneless pork, with a 27% reduction in environmental impact. Finally, cod is more environmentally friendly than plaice (− 49%) and swordfish (− 75%).

In the process of improving foodservice from a sustainability perspective, the role of digital technologies must not be forgotten. A comprehensive concept of "data-driven sustainabiliy" should be used to achieve environmentally friendly benefits and competitiveness in foodservice. Artificial intelligence technologies can be used to track food consumption patterns, identify food waste quantities and composition, and provide areas of intervention without forgetting the importance of human expertise. Food waste measurement should be based on real-time monitoring systems that continuously analyze food consumption and waste, by relying on digital scales or scanners. In this view, foodservice operators should use digital technologies to reduce food waste, such as machine learning, artificial intelligence, image recognition, natural language processing, the Internet of Things (IoT), blockchain, and reinforcement learning. In the healthcare sector, these monitoring needs through digital technologies are multiplied by the different interests at stake: economic, for reducing costs linked to the supply of resources and waste disposal; and social, for making sure patients are treated fairly, increasing consumer satisfaction, and reducing the level of malnutrition.

Index

C. Bux, *Strategies for a Sustainable and Environmentally Friendly Hospital Foodservice Management*, Environmental Humanities: Transformation, Governance, Ethics, Law 6,
https://doi.org/10.1007/978-3-031-82050-2

Kolibri 360 GmbH
Ferdinand-Jühlke-Straße 7
99095 Erfurt, Deutschland
produktsicherheit@kolibri360.de